A Phenomena-Based Physics

Sound ~ Light ~ Heat

Volume 2 of 3
for

Grade 7

Manfred von Mackensen

Sound ॐ Light ॐ Heat

Electricity, Magnetism & Electromagnetism,
Mechanics, Hydraulics and Aerodynamics
as Subjects in an
Introductory Physics
6th through 8th Grades

Volume 2 of 3
for

Grade 7

by
Manfred von Mackensen
Pedagogical Research Department
Union of German Waldorf Schools
Kassel, West Germany, 1992

translated and edited by John H. Petering
August 1994

Printed with support from the Waldorf Curriculum Fund

Published by:
Waldorf Publications at the
Research Institute for Waldorf Education
351 Fairview Avenue, Suite 625
Hudson, NY 12534

Title: *A Phenomena-Based Physics: Sound, Light, Heat: Grade 7*
Author: Manfred von Mackensen
Editors: David Mitchell and John Petering
Translation: John Petering

German original copyright 1987, 1992 by Manfred von Mackensen,
Kassel, West Germany
First printing © 1994 by AWSNA Publications

Revised edition © 2018
ISBN 978-1-943582-25-9
Layout: Ann Erwin
Proofreader: Ruth Riegel
Cover: Image from the *Brockhaus and Efron Encyclopedic Dictionary* (1890–1907) in the public domain

Translation unrevised by the author.
All responsibility for the translation rests with the translator.

Permission granted for use by the English-speaking Waldorf teaching community.
All other rights reserved.

All rights reserved. No part of this publication may be reproduced in any form without the prior written permission from the publisher, except for brief excerpts.

Table of Contents

INTRODUCTION

The Waldorf Approach	7
Teaching, Technology and Life	8
The Landscape of Science	10
Alternative Science – No Thanks?	14
Developmental Psychology	16
Pedagogical Methods	19
To the Teacher	20

GRADE 7 OVERVIEW ... 26

GRADE 7 ACOUSTICS

The Tuning Fork	27
Oscillation and Frequency	29
Experiments in Acoustics (Ac)	32

GRADE 7 OPTICS

Introduction	37
Mirror Phenomena	38
The Structure of the Image with Mirrors	48
Physics in School	57
Steps to Understanding the Camera Image	59
Experiments in Optics (Op)	62

GRADE 7 HEAT

Introduction	67
Warmth and Surroundings	67
Thermal Expansion	69
Thermometers	70
New Sorts of Ice Experiments	71
Experiments in Heat (He)	73

Contents

Grade 7 Electricity

Introduction	77
Galvanic Electricity	78
Volta's Pile	81
Voltage and Current	84
Current Strength and Short Circuits	86
Voltage and Resistance	88
The Experience of Electricity	90
Experiments in Electricity (El)	93

Grade 7 Magnetism

Introduction	98
Experiments in Magnetism (Mag)	103

Grade 7 Mechanics

Overview	106
The Lever	106
Wheel and Axle	109
About Fixed Pulleys and Block & Tackle	111
Inclined Plane, Screw and Wedge	113
Experiments in Mechanics	114

Endnotes ... 118

Selected Bibliography ... 120

Introduction

I. THE WALDORF APPROACH

The following chapters describe a selected area of teaching—physics—the contents, initial thoughts, yes, even the entire construction of the teaching material and many teaching methods, all drawn from the treasure chest of Waldorf pedagogy. But, simply by working through the descriptions for an *individual* physics topic given here, someone foreign to the Waldorf methodology can hardly achieve intimacy with the whole of this pedagogy. And yet, it is only from this whole that such particulars (e.g., the physics lessons) can arise or be available.

What is presented here can be only a kind of *indication* of this whole, that at first stands open and unknown. At the same time, this material *can* serve as a stimulus also for the teaching efforts of those not familiar with Waldorf education. The present material could even be useful from a variety of viewpoints, e.g., finding little-known physics experiments.

Still, in spite of these obstacles, the fundamental concepts of the Waldorf approach to natural science should reach a wider audience. Certainly, it will be in the spirit of the *preconceptions of the author* and the working-group here in Kassel (the basis for this whole project), including their understanding about scientific theories, developmental psychology and commercial starting points—insofar as an introduction can draw comparisons to contemporary ideas.

Although this book alone is not really an introduction to the essentials of Waldorf pedagogy, nor can it even fully develop a scientific foundation for our approach, we must attempt to highlight our preliminary understanding, which has been used effectively, and draw attention to it in the context of the contemporary science education literature. The starting points of our work should be made visible; thereby, skeptical readers will be able to select something useful, try it out, and perhaps have it make sense sooner than later.

Whoever is new to this subject, or initially wants to spare themselves a discussion of foundational scientific theory and educational psychology (possibly many readers), could perhaps read only the final section of this introductory chapter: "VII. To the Teacher," which summarizes many theoretical and practical points. After that, they could experiment with the various lesson outlines of a phenomena-based approach to physics. Thus, they will have the necessary examples in front of them, and later may go back to read the rest of the background to this approach to studying the sciences, especially physics.

[Note: The following sections have been summarized from von Mackensen's detailed discussion, as most of the educational and philosophical authors cited are not well-known in English-speaking countries. For a full critique of the philosophical underpinnings presented by Dr. Mackensen, a full translation of his introduction is available from the Waldorf Publications clearinghouse upon request, at no cost.]

II. TEACHING, TECHNOLOGY & LIFE

Basically, we have less and less success guiding young people in learning and work, both at school and at home. Education slips away from this generation; although the amount of scientific work given to them is greater than ever before! Elaborate reforms, allocation of financial resources and dedicated administration in the schools have not helped the whole. Plainly, neither the analytical, technical, nor legal modes of thinking truly relate to those deeper aspects of the human being capable of transformation and learning. The human forces that allow people to form their spirituality, enliven their soul, and enable them to grow plainly lie in another realm from the modern scientific way of thinking.

Educational writers themselves point to a need for a reorientation of science teaching to make it relevant to people's daily lives, in view of the "lack of impact of a technically-oriented science teaching, which tries to tackle the teaching and learning process using the [quantitative] methods invented by that same science...." (Duit 1981) The fact that schools aren't easily reformable with technocratic means is due to the fact that

> the first phase of reform was ... too foreign from the children, went off into abstract scientific criteria. ... Along with a structurally-rooted, growing unemployment, the principle of competition shifts down into ever lower grades. And, as schools allow such selective and competitive modes to creep in, it only increases the impoverishment of students in terms of social experiences, emotional connections, and feeling-filled learning possibilities.
>
> On the heels of the discovery of possibilities for early learning came performance-based teaching programs, minutely planned class periods (with well-planned teaching behavior and expected student behavior), and increased *demands for more content* and more technical-professional scientific topics in the lessons. To our dismay today, we actually *establish the isolation and alienation of the human being."* (Silkenbeumer 1981)

Perhaps we can explore further this contrast between the bright, well-defined world of science with the dark, less consciously understood world of life, albeit in a living manner, using open-ended characterizations, and always mindful of the whole range of human activity.

Another description of how these two principles from the world of life and of science infiltrate everything is given in the last great letters of Erich Fromm. He characterizes them as "having" and "being": "Indeed, to one for whom having is the main form of relatedness to the world, ideas that can not easily be pinned down (penned down) are frightening—like everything else that grows, changes, and thus is not controllable." (Fromm 1976) After showing how the current crises of humanity spring from the ideology of fixation and having, Fromm describes how the modern technological world promotes a marketing-character, people who "experience themselves as a commodity ... whose success depends largely on how well they sell themselves on the market" and, dedicated to pursuing the quantitative side of life, become *estranged* from their living surroundings and from their own self. "Most striking, at first glance, is that Man has made himself into a god because he has acquired the technical capacity for a second creation of the world, replacing the first creation by the God of traditional religion. ... Human beings, in the state of their greatest real impotence, in connection with science and technology *imagine* themselves to be *omnipotent*." (Fromm 1976a)

If we shy away from Fromm's pedagogical indications, we can still value his observations and views and note again that *it matters for the whole of humanity, for everyone*. So, whether it admits it or not, education also is tensioned between these poles of "having" and "being," or between convergent and divergent issues. (Schumacher 1979)

Convergent issues are of a type which *can* be dealt with and converge on a stable solution simply by applying more time, intelligence, or resources (even without higher forces of life, a wider consciousness, or personal experience). Dealing with *divergent* problems with more study only produces more polarity— they require a higher level of being, beyond mere logic, where the seeming opposition of polarities can be transcended. A classic example is finding the right educational methods, where the right formula or answer is seemingly never found in the opposing ideas of more form vs. more autonomy. Divergent issues involve a higher realm, including both freedom and direct inner experience.

We all must cope with our scientific world; its development is necessary for a free consciousness. But now it challenges us to find its limits so that the living environment, society, and the human spirit can breathe. **This problem is truly**

Introduction

the deepest, most global problem of our time. Within science, challenges lead us to new *scientific theories,* and many, diverse initiatives exist, but with only meager results up to now. (We will indicate our relation to each of these initiatives below.) When our thinking within the living world is challenged, it can lead to *Goetheanism,* to a phenomenology rich in living ideas. But, mere concerns in the living world lead only to mythology.

III. THE LANDSCAPE OF SCIENCE

PRESUMED CERTAINTY VS. LIFE

In recent years there have been many articles published that discuss the philosophy of science (How do we know the world?). Science teachers adopt the same well-accepted, positivistic habits of thought as more technical writers, presuming that conventional quantitative methods do give us a real knowledge of nature. Yet, if we delve more deeply into what many writers in the philosophy of science say about this, what emerges is not at all so cut and dried—nor clear. The fundamental question is: What is the relationship between our selves and the world out there? This is one of *the* fundamental questions of philosophy. Many philosophers such as Descartes, Kant and Goethe have wrestled with the question. Does my mind (soul, spirit) have any real connection with my body (the world of matter)?

Even contemporary educators acknowledge the subtleties of this relationship: "What scientific research achieves, in the best case, is not a plain portrayal of reality. The theory of science has shown that the researcher is not only interwoven with his experiments and the measuring apparatus he uses, but also (in a much more subtle but potent way) is intermeshed with his whole theoretical and mental framework, with his very concepts and way of questioning, with his definitions and hypotheses."[1] (Wieland 1981) And in 20th-century physics itself, exact results have come about that do not illuminate a clear reality, and indeed never can. The Copenhagen Interpretation about quantum theory (1926) early on led a leading physicist to say "... the reality is different, it depends on whether we observe or not ... and we must remember that what we observe is not Nature itself, but nature *as expressed to our manner of questioning."* (Heisenberg 1959; see also K.F. von Weizsäcker 1971)

This awareness, beginning first in physics, now spreading to many other fields, forces us to recognize that our relation to the world is not as simple as we might have thought. Although there are myriad ways of considering nature, for *our* path the image of nature we develop fully develops and ripens only *through the human being*—the human being is not superfluous. Our preconceptions (what T. Kuhn calls our paradigm or framework of what we take the "world" to be) matter a lot—they condition the very character of what we come up with. If we really think about it, the very reality which science presumes cannot be thought about as something real "out there," with ourselves as something separate, a kind of onlooker, this reality consists in our mutual relationship and intimate connections with the world.

THE SOLE AUTHORITY?

While it is true that the conventional literature doesn't claim to be the only valid method, at least not *explicitly,* still, this attitude is *implemented,* in that empirical or positivistic habits are accepted and implicit in the way science and teaching are pursued. Actually, some educators have come very close to touching on these questions, and their ideas are significant not because they pose a particular problem, but because they discuss *standards* of what is "good" science.

They usually presume that our knowledge and theories relate to a pre-existing objectively material world. And "what gives [this method] certainty is a clearly marked path back to the evidence of data." (Jung 1982) This presumes an aspect of phenomena separate from our conceptualizing, which is part of an objective world, existing *before* we think about the phenomena. So "reality" must consist in a *collection of basic phenomena* (i.e., all our perceptions in the living world) which is accessible to all persons.

So the fundamental question becomes: Is there some sort of pre-existing objectively material world "out there" and our observations lead us "in here" (in my mind) to think about it and formulate theories and understanding? Are "phenomena" *something* independent of me, pre-existing, or are phenomena actually my experience of a relationship between self and world, awakening and coming to light in the perceiving, beholding human being? This is a very tricky, subtle but *very* important question.

STRENGTHS OF MODERN SCIENCE

On the other hand, in a pragmatic view, people set aside the questions of whether the knowledge is "real," and presume the task of natural science is to simply reproduce an image of this layer of reality as precisely as possible, and organize it in a comprehensible way. So, even if a subjective prior conception and historically related world paradigm play a role, then the image we build up (modern natural science) would have the following strengths:

- The path from theory back to phenomena can be clearly retraced using understandable concepts, as must be since this method deals only with the aspect of phenomena understandable in thinkable, quantitative categories. Thus, this whole construct will be *comprehensible.*
- As we go beyond mere acceptance of an invention and mere application to build up theories of immense breadth at the highest level, daring and bold intuition plays a role. Such ideas open unbounded possibilities of activity and schooling to the human intellect. Thus, the whole is *inspiring.*
- In applying such theories and methods in working with nature, technology arises; we needn't argue its effectiveness. Such a science is *usable.*

Even if the whole range of scientific theories isn't taken as objective general truth, nevertheless most people feel that, in fact, it *does* achieve the best that could be achieved *by* and *for* people, with the means at their disposal. Although it might encompass and weave in more of reality with ever better revisions of its theories, they can't see how anyone could achieve something more understandable, more inspiring, or more effective! Most people concede such a natural science may not be the only one possible, but believe it is the best that can be achieved, and thus should be acknowledged and supported by everyone.

PURE PERCEPTION, A WELL WITHOUT A PIPE

We have explored the conventional ideas in order to contrast our approach of phenomenology or Goetheanism. Our starting point for all knowledge and thinking is actually *perception* through the senses, the active participation and perception through human senses of a living human body. But pure perception is very elusive; we can't actually *say* anything about the senses unless we go beyond pure sensing. We can explore the world by means of perceptions, but we can't discuss the world in perceptions. Nevertheless, we start with perception as a relationship, one that goes far deeper than these concepts we are using to discuss it, and even lies *prior* to them.

But, we can't prove this, since proof already goes beyond perception to use conceptualization. We can have pure perception only when we *hold back* our capacity for forming ideas and concepts. (Steiner 1918) **Only pure perception exists per se—but we cannot say anything about it.** The reality of the way it arises in the human soul is actually something holy; what we could say about the content of perception is that it is a human product, always a prior conception and a theory conditioned by the living world.

THINKING REVIVES IN PERCEPTION

A person could object: Then every theory is just arbitrary; but if they are neatly formulated, they relate to the world because they are technically usable. We respond: True. Concepts and theories are certainly added to perception by *human* effort, however they plainly prove to belong to the world if people have made an effort, but correspond to experiences of *a different* side of the world, one not revealed in perceptions. Our thinking is an essential complement to pure perception. In thinking, people have a kind of organ for this other side of the world, which is most active when we reflect on the world of senses rather than the world of thoughts. Goethe, in contrast to many modern thinkers, understood perceiving as a particular conceptual activity.

PARADIGMS AND THE LIVING WORLD

As soon as one transcends this reductionist way of asking questions, and thoughtfully begins to work with the qualitative side of human experience, there is no longer such a thing as phenomena *independent* of theory (conceptually organized perceptions). The phenomena are the aspects of the world we live in through being conscious in perception. In the process of thinking, through a balanced consciousness, out of the experience of perceiving naturally unfolds a definite kind of conceptual activity. The more we live in the perceiving side of this balance, the more the true being of the world will actually speak in the thoughts and concepts that the phenomena lead us to form. Pure perception exists per se; phenomena are already passing over to a more conceptual side.

This is important because the implications are very broad and deep. The reductionist approach, considering mind and self as separate from the world out there, has a split implicit in it: "The spiritual and moral ... dealt with as epi-phenomena. ... On the one side is human life with its values and choices, on the other, a scientific Utopia: Man as Machine." (Jung 1981c)

INDIVIDUAL DECISIONS

Since we can prove neither method with logic alone, it must remain a free decision of each person which way they wish to understand their relation to the world: theories about the world that allow one to manipulate the world, or a relationship of perceiving. Basically, we can only say that the author has worked through the phenomena in this way, and *intends* to work with them thus. The Western tradition tends toward manipulation of the nature out there; a complementary approach lives more richly in the phenomena and uses more the subjective, qualitative side of perceiving. It does require a carefully schooled, balanced, thoughtful consideration of the phenomena and of our process of forming concepts.

IV. ALTERNATIVE SCIENCE – NO THANKS?

LONGING AND HORROR

We have reached the most explosive place in our introductory discussion. Even in one of the most inviting works on teaching methods, it can happen that when it comes to the theme of alternative approaches to science, the whole style changes dramatically: Irony, dire predictions and all-out polemics surface. This *although* or perhaps *just because* alternative science increasingly becomes the vague object of longing of many people.

The editor of *Chemistry and Technology News* proposes, "Do you believe that it would be possible ... to develop a rational scientific method which doesn't pursue just this reductionist approach to nature, rather a method which sees as much of the whole and simultaneously isn't useful technically? You say yourself that's a royal task. ... Couldn't we at least strengthen a consciousness that there exists a natural-scientific-technical world conception other than our own? Are there certain sciences which are not intended for utility, but earlier on proceed in a phenomenological way, and if there are, shouldn't we pursue them?" There is nothing to say but YES!

Another author expresses alarm that "science is not features of rationality or freedom, not basics of education; it is a commodity. The scientist himself becomes a salesman for this merchandise, they are not judges about truth and falsehood. ... The starting point is not truth, or the newest level of science, or some other empty generality; the starting point is the equivalence of all traditions." (Feyerabend 1981)

The anti-science effect, which has grown so strong today, arises not only because of ecological, political-social crises, but is due to a split in the living stratum of the soul into calculable scientific concepts on one layer and feelings and moral (or animal) impulses on another, which directs practical dealings in life, and yet becomes ever more isolated. It is also on the rise since people no longer hesitate to speak about this split out of "respect" for the acknowledged science.

Introduction

HUMAN POWERS

In the last centuries the scientific view of nature has established only what the Cartesian revolution, the methods of Galileo, Newton, etc., contain: a kind of imperialistic knowledge that serves the orientation of the human being toward control of the cosmos. The reason the causal-analytic approach to natural science requires an alternative is not because it provides too little *information* about the world, but rather because it doesn't activate our soul-spiritual forces.

WEAKNESSES OF PHENOMENOLOGY

According to the above ideas, there are three things phenomenology cannot have: It is not understandable in laymen's terms (in concepts that are definable in a mechanical framework); it has not progressed very far nor inwardly developed (doesn't have 400 years behind it); it is not usable in a naïvely unconscious way, in a mechanical sense, without a look-back at the whole situation in nature.

The most negative is a lack of external provability, without a mechanistic mode of thinking. This is unavoidable, since this provability arises from just those conceptual tools that now should be extended and complemented, namely the quantitative approach, and ultimately mathematics.

EXTENSION OF BOTH METHODS: PROPADEUTICS (LIBERAL-ARTS INSTRUCTION)

Alternative and rule-based science complement each other, and neither possesses all the necessary aspects for education; we can dispense with neither one. Even where we can explore with a quantitative approach, we must go inward.

For example, in the gas laws, if we wish to really penetrate the matter, we must go inside the mere formula $V_t = V_0 (1 + 1/273\, t)$. As the heating is increased, the increased tension in the gas container (pressure) is experienced as a reflection of the activity of heat and the increase in measurable volume.

Analytical thinking initially creates distance, but then allows an inner connection with the processes in the world, insofar as the magnitudes in the formula we work out are inwardly felt and experienced. Such a going-inward is the starting point of all phenomenological research.

If the advanced technical material presented at the introductory level of a broad, liberal-arts curriculum should be limited, what should be correspondingly expanded? This leads us to a human basis for a new teaching method: basing the themes on the development of the student. The questions then become: In which grade should physics start? What ideas should it pursue?

V. DEVELOPMENTAL PSYCHOLOGY

TRADITIONAL DIVISION OF PHYSICS MATERIAL

In the pedagogical works of Rudolf Steiner there are many places where he points to the essential reorientation of the lessons at the onset of pre-puberty. He indicates how, mainly after 11-2/3 year (6th grade), the young person separates out the living and ensouled from the dead aspect in the surroundings and is now able to recognize it. Therefore, we can be attentive to this moment, understand why the study of mineralogy and physics can then begin, and relate to the young person's developing natural interest in life.

CURRICULUM OVERVIEW

In 6th grade—when the physics studies begin—the way of regarding things is still phenomenological and imagistic in all the topics. It is not yet abstracted in the sense, for example, of deriving general Laws of Nature. The material-causal method of school physics is kept out. However, in the **7th grade** this comparative-imagistic approach already receives a new direction. Certainly not toward scientific models, atoms and the like, but rather toward work, livelihood, trade connections, and thereby to technical applications in life. Since young people will at puberty begin to distance themselves from their parental home, they are occupied with the question: How can a person help himself along in the world through clear ideas? And, underlying this: How can a person contribute something of value in the outer world of work?

At the end of the 7th-grade physics studies, *as* an essentially technical topic, we present the mechanical theories involved with the use of levers. Simple experiments, scientific systematics and a technical-practical understanding of the apparatus all occur together in mechanics. Out of such a treatment, we reach the starting point of classical physics. However, a systematic treatment of the other topics in physics founded on this does not quite follow yet. Even in **8th grade**, we will use quantitative or expressible formulas only in particular cases—for example, in the treatment of current in a circuit or in the pressure calculations for fluid mechanics of air or water—but they will still be connected with direct observations in the classroom.

In 9th grade after the material-causal explanations of the telephone and locomotive have been treated thoroughly (for example, with current-time diagrams, vapor-pressure curves and thermal mass comparisons), this phenomena-based method arrives at a quantitative systematics for the first time only in the **10th grade.** There, even while quantitative, the view shifts back to the human being: How we predict through calculations of a parabolic trajectory is now thoughtfully considered as a phenomenon of knowledge. And, only after the question of perception and reality in the "supra-sensible" is worked through in the **11th grade** via the study of modern electrical inventions (Tesla coil, Roentgen rays, radioactivity), do the light and color studies of the block in **12th grade** penetrate again to a Goetheanism. However, the methodology [and epistemology] is now clear to the students.

So, moving toward the 9th grade, the curriculum increasingly lays aside the phenomenological approach, as we increasingly lay aside nature. With the analytical method and use of technical instruments, a non-spiritual, material causality chain now becomes the goal of the lessons. This is what is appropriate here and therefore—pedagogically viewed—nonetheless a Goethean method (Goetheanism in the consideration of a thought sequence). From these technical-practical topics of the 8th and 9th grades and their social implications, we turn around again in the 10th grade and focus on the thinking human being. The process by which we develop knowledge and the relationship it has thereby to the world is raised to the phenomenon here. Through such a careful and interconnected arrangement of steps in physics, the young students gradually awake to an exploration of their own methods of knowing—and that is the real objective.

SUMMARY

The so-called abstract study of natural science, using causal-analytical methods, actually lies near other paths (see chart next page). First we can consider the overall experience reading downward in the central column: A restriction to increasing abstraction and models, providing the all-permeating meaning of the world, and degrading the initial holistic experience to a misleading reflex, a byproduct of the model (an illusion). Knowledge becomes imprisoned in the lower circle, and people run around in a senseless experience of a causal-analytic mechanistic world. They develop gigantic technical works and enjoy everything that they love.

In contrast, working upward on the right is a deepening of a phenomenological way of considering the world, leading people to truly grasp who they are, and thereby building a basis for a new approach to the spirit—a holistic experience. They enjoy what they *realize*. With this we note how the 6th grade begins with holistic knowledge; the 9th grade moves to causal-analytic thinking but not yet models; and mechanical thought in 10th, where we stop.

Introduction

TWO PATHS OF NATURAL SCIENCE CONTRASTED

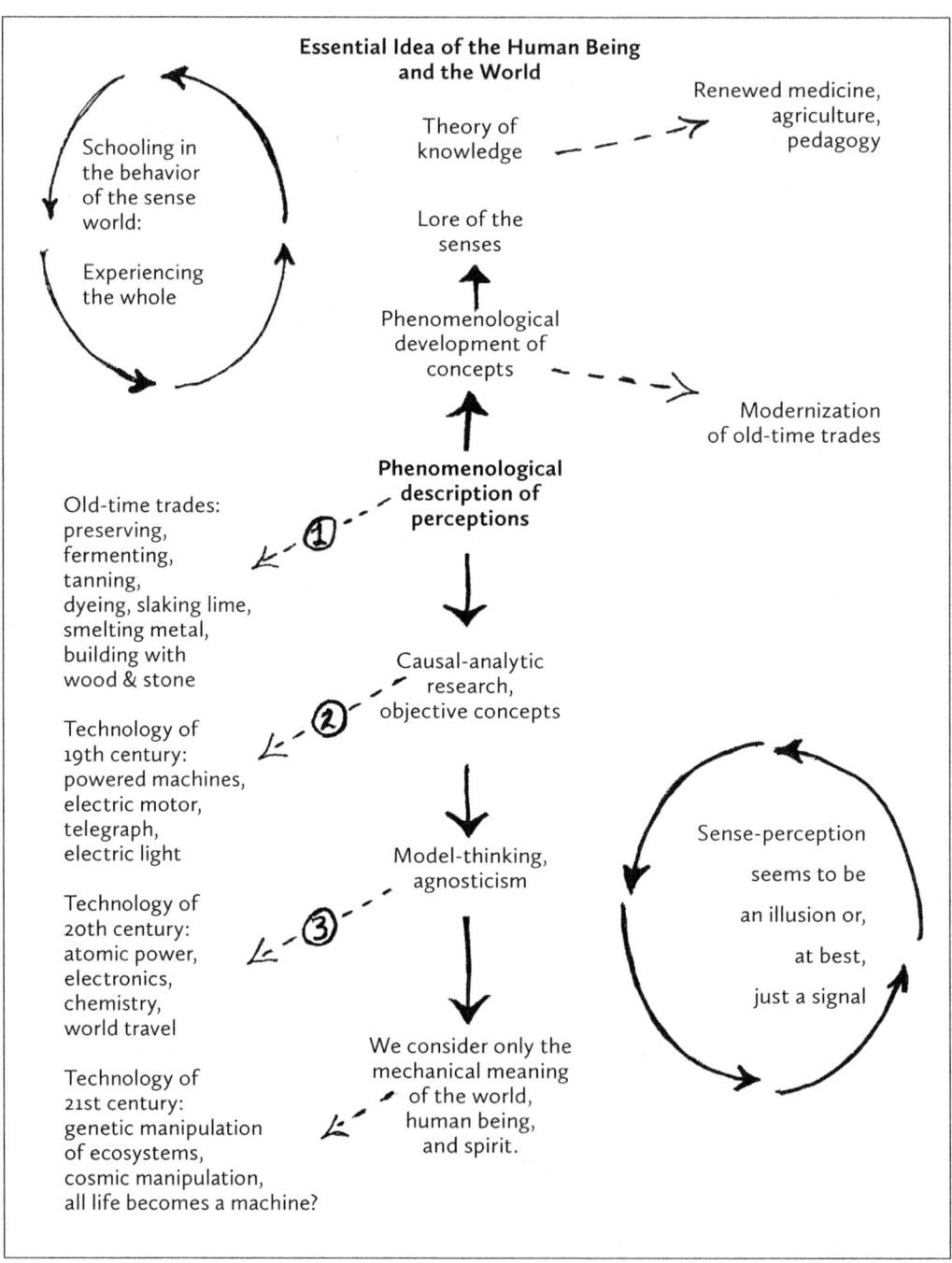

VI. PEDAGOGICAL METHODS

Nowadays, natural science is pursued with the goal of bringing facts to light and utilizing them technically. Whether research makes an impression on the public depends on the usefulness of these facts and on what can be produced from them. The analytical method is impressive and shows the basic significance of the things investigated. But, how do things stand with knowing?

In order to unify and order the innumerable facts, physicists build theories and conceptual models, e.g., the wave model of light. They know, in principle, that every model has limitations and that, above all, there are always phenomena that will contradict it. A model applies only to certain phenomena; for others, one must construct a contrasting model for the same topic. So it is that, admittedly, no knowledge of the whole arises. Nevertheless, scientific models such as the light theories or the atom-concept (which never touch the essence of the matter) are commonly viewed as the essence, or at least as steps to the essentials—as much in the popular view as in scientific consciousness. In thinking this, we overlook the fact that questions about "essence" fall entirely outside the categories of physics, cannot be included in them, and equally cannot be answered by them. Nevertheless, to the student every model is naturally taken as an answer to essential questions, a statement about the ground of existence, although the students are not consciously asking about it! The narrowness of these kinds of answers cannot really be appreciated by the student, since s/he does not yet comprehend the role played by pure thinking in forming world-conceptions, or the way a free activity of the human being is connected with it. The student is oriented more to perception.

Abstract conceptions, however, which do not originate from the phenomena, become misconceptions: as if not invented by us, but rather as coarsely material or also as magical beings which we light upon and then believe that they stand as final cause behind everything there is. And so the phenomena give rise to an ontological misunderstanding of physics[2] [misunderstanding the nature of being or reality].

Goethe's natural-scientific work, as set forth by Rudolf Steiner, developed a little-known qualitative way of observing nature and now is constantly producing and extending a literature of Goetheanism and phenomenology. This concrete path built around an intercourse with qualities of perception will actually bring the perceiving human being to reality. Due to the style of teaching typical of the later elementary years, there is a fundamental need for qualitative natural-scientific methods, as only these make an attempt to work in accord with the essentials, to work on just those questions deeply experienced at this stage of life. The question of whether such an approach promotes complete understanding can remain open, but certainly it is more complete than that promoted by models. For us, the pedagogical value of the teacher's striving and his inner, spiritual activity exceeds the value of mere knowing.

TWO PATHS TO NATURE

This suggests two methods of conceiving of nature (and thereby also the world) which at first are unconnected, namely: (1) the material-causal method, which wants to find a model of the ultimate cause through analyzing matter, which is thought to lie at the foundation, and (2) the Goethean method of phenomenology, which seeks to order rich perceptions of nature as a totality. These two ways of approaching nature are both contained in Dr. Steiner's curriculum. However, he didn't intermix them, nor did he present them interspersed with one another. Rather, we find them organized into the curriculum with the analytical following the phenomenological—the aim being to place them in an evaluable contrast by the 12th grade.

VII. TO THE TEACHER

HEALTHY AND UNHEALTHY PREPARATION

This book has been written precisely for those who won't receive it first: the students. Teachers will doubtless receive it, perhaps with a sigh of relief. It may seem that in it lies a wealth of physics material, neatly packaged and awaiting use like pre-assembled concepts, stashed away in a deep freeze. The impression that I have carefully tried to create, to be understood perhaps only later, is that such a hope cannot be realized through a book; and if it could it should be forbidden! The students do not live off the stuff from a book, but from the initiative of the teachers and from their spiritual wrestling and persistence to blaze a path for them on how to see the world. As the teacher learns, so the students learn. Whoever is in good command of the subject—knowing that she must not only freshly order things but must unlock new sides of it *for herself*—will be able to teach out of direct, spiritual experience—she will inwardly reach the class. In order for the physics covered here to encompass this dimension, it was necessary on the one hand to make the subject easy for the teacher, so it could be expressed in words, and on the other hand, also necessary to make it hard for the teacher. This challenge comes from me, since I can understand the full scope of the material only in certain scientific and pedagogical terms.

Following Goethe's beginnings and in the spirit of Dr. Steiner's theory of knowledge, a person should *attend to nature's phenomena and simultaneously attend to the inner activity and experiences of the [perceiving] human being;* then the person must attempt to break free of the rigid [mental] labels of "particles," "waves" and "energies." New, basic ideas are of great value to go more deeply into nature, but these are radically different from the traditional concepts of school physics, and from a materially disposed, conventional understanding.

My ideal would be to set aside the quick definitions in phoronomic terms [purely mechanical motion], and lead everything over into open, seemingly indeterminate concepts. What light "is," what sound is "in actuality," what is "the basis" of heat—all this cannot be clarified without further ado, but must be achieved by experienced observations and thinking on the part of the teacher and the students. We call upon the reader's own perceptive judgment. (Compare Goethe's concept: *Anschauende Urteilskraft,* "perceptive power of beholding.")

Clearly physics classes shouldn't be merely a list of points out of the great book of scientific authority, but should be spontaneous, without limits at the outset to questions involving only quantifiable measurements. Naturally, this requires a lot, namely: **an alternative way of thinking,** the greatest possible individual sense of orientation, and **a familiarity with the phenomena.** Only to the extent that this text seeks to pursue such difficult goals, should it be allowed to appear. An easy, practical collection of obvious school-knowledge—provided it is good—is only a help to the teacher who is already in desperate straits; however, it undervalues just these deeper needs of the students.

Aside from aiding such a direct, untrammeled, personal spiritual quest for knowledge, teaching has, naturally, a more ordinary side: We also must furnish the students with actual knowledge of the thinking and the day-to-day business of our technological civilization. For this the teacher needs information that is refined and prepared. So I have interwoven a few such indications in the main text and also provided some related topics. There we can see how we can work without abstract scientific models and hypotheses so far removed from the phenomena.

EXPERIENCES WITH THIS METHOD

Interesting teaching experiences have arisen meanwhile from colleagues who have received this book just because physics was new for them. In the worst cases, the students said "That was, certainly, all very logical," meaning that the phenomena were already clear (they weren't led to previously unknown phenomena). On the one hand, they were caught up too little by the wonder of the world; on the other, not enough difficult questions were given them.

Introduction

Now, this raises several issues. The students have a point: New facets of the world must be shown in wonder-filled, previously unknown effects. The students like to plunge into manifestations of the lively activity of strange forces, they like to see things that can be conjured forth with simple manipulation. Whenever the teacher surprises them with unexpected phenomena, that makes an impression! The manipulative, perceptible side of the world must come to the fore. True, without a wakeful musing and reflecting over the experiments, a certain superficiality and inconsiderateness would be fostered. Nevertheless, I do a few direct, surprising presentations. But, in caring for the perceptive side of things, one should not neglect the thought-aspect.

To an art of experimentation belongs also an art of conceptualization. A person will place himself perceptively into everyday phenomena and, out of his own careful pondering, form new concepts, so as to build wholly on the phenomena. Then experiments that appear commonplace will no longer seem merely logical but, perceiving with new eyes, will allow the thinking to reach up to the working ideas of nature so they really speak (see Goethe's essay on plants, regarding his "seeing" the archetypal plant). It shouldn't happen—as it often does in conventional physics—that the teacher simply presents how things are supposed to be seen. Goetheanism as pronouncement remains mere semantics.

Conventional physics, on the one hand, relies on everyday thought-forms, which have a life of their own and therefore easily take root in the students. On the other hand, the more the teacher has the foundations of Goetheanism in herself, still utilizing and experiencing it even on days when she doesn't teach it, the more she is able to call it forth out of the simplest phenomena and give it a permanent basis in the students' souls.

Using a book to show the way to do this is, naturally, risky. In the worst situations, the impression could appear that the proposed material is too childish and that the 6th-grade material should already have been done in the 5th grade as it lacks causal thinking. By all means, twelve-year-old students (grade 7) should already practice a thinking based in the compelling links or relationships between specific phenomena. We only have to try and develop a thinking *freer* than a thinking that explains the appearances out of processes of the matter, which is supposed to make up the substratum of reality. This will be shown particularly in the section on light.

Introduction

UNUSUAL LANGUAGE

It is necessary in the 6th and 7th grades, and somewhat in the 8th, that the students immerse themselves in the qualitative aspect of things. Thereby, their experience receives an orientation toward objectivity, not through analytical abstractions, but in careful expression and in connecting to the phenomena. If one wants to work in this way, one needs *expressions* and *a language* that embraces *the feeling side of experience.* The one-sided cultivation of the cognitive faculties—basically only rational instructional elements and goals only in the cognitive domain—fails to take into account and neglects the feeling and doing side of the human being. The student learns individual pieces of knowledge. But the feelings will not be expressed, developed and differentiated: "The feeling of senseless learning becomes often a feeling of senseless life, the seed of a depressive orientation." (Fischer-Wasels 1977) Through abstract teaching, the desire for an emotional reality can lead to a splitting of personality. As in many such articles, a rigorously restricted subject matter is called for and, for the natural sciences, a strong consideration of the phenomena. [See in the U.S., the move by the PSSC curriculum project to emphasize simple phenomena with simple apparati – Trans.]

As experience shows, such a *restriction* of the science content will be of little help if a *new art of thought*—a *different language* and an experiential way of working connected with the phenomena—does not simultaneously replace a mere taking up of one phenomenon after another. One can only put *something else,* not nothing, in the place of an abstract thinking that has become a scientizing. This "something different" should engage the complete *experience* of the human being, not only our *conceptualizing* faculties. That is to say: In the course content, new fundamental concepts are necessary! The technical language doesn't suffice.

What is attempted in this book is to utilize such language as *unites* the feeling with the conceptual. But, this is the main snag for professionally trained people. Rather than an illumination of actual perceptions, they expect to combine definitions and concepts purified of every experiential aspect (e.g., *phoronomy* or kinematics; motion without regard for the actual units). Already on pedagogical grounds, we cannot here concur with this goal. Thus, we must attempt to intersperse these rational and emotional instructional elements, not only like a spice, but rather to draw them right out of *the things at hand.* From the most fundamental perceptions to the most all-encompassing thoughts, it is important to always speak to the whole human being and to truly look on concrete phenomena of the world right through to the end. That leads necessarily to unconventional, open-ended concepts and indications.

Introduction

We take into consideration that these new, open-ended fundamental concepts have their source in perceptions saturated with feeling, they lead on to meaningful thoughts, and they have a structure that can be evaluated as a result of their relationship to actual perceptions. We aren't dealing with a complex of feelings arising from some kind of subjective evaluation, but rather with elements of science [a conscious process]. The full earnestness of the task of physics teaching, as presented herein, will work as a stimulus when one has the insight that the central concepts we strive for aim not at a childish prettying-up of the barren subject matter of nature with all sorts of subjectivity; rather our aim is to open up a new scientific reality out of our own personal perceptions—however modest the beginning introduced here might be.

You might notice that the new concepts in the section on light are a bit further developed toward a Goethean phenomenology. With heat, especially in grade 6, such a Goethean approach would be impressive but also becomes more difficult to grasp. Acoustics in grade 6 is also organized around unconventional ideas, and only in the subsequent grades will become more familiar. But in acoustics, these new concepts do not take such an unusual form as in light studies. Electricity and magnetism are presented in a qualitative way and lead to concepts formed in a particular way, though these remain mainly empty forms, as yet unfilled with much that is compelling or germinal. In mechanics, least of all have we made a beginning into an experientially clear treatment of force and pressure (hydromechanics).

We are dealing here with a search for an understanding that takes as its starting point the human being. In order to live into such a new mode of approach, the epistemological orientation must be spoken of in an extensive work, amply supported by philosophical and anthropological points as are here only begun. And the physicist would nevertheless still not be free of a deep uneasiness—for, instead of studying the rewarding scientific developments of this century, we have apparently given them up in favor of a new consideration of Aristotle. Whether this is necessary, something to put up with, or just foolish cannot be determined from concepts but only in one's own intimate meeting with the perceptions.

If I occasionally draw on concepts from the writings of Dr. Steiner, this should provide those familiar with his work a quick means for further reading. However, the seeds of the phenomenological method and what is therefore useful in teaching are not dependent on these references. Thus, I hope also to invite many useful contributions from the non-Waldorf teacher.

THE BOOK'S LAYOUT AND ABBREVIATING THE MATERIAL

Concerning the structure of this series of three books for grades 6, 7 and 8, may I say that it follows the sequence of grades 6 through 8. The experimental descriptions are, with some exceptions, extracted from the text and inserted after individual chapters. Appended in Book 3 is a list of equipment and, as a possible deepening, a few supplemental topics.

The extent of material presented often exceeds what can be handled in one block, especially if a teacher is presenting this for the first time and isn't at all familiar with many of the phenomena. However, one should not omit an entire topic, as for example, the one usually coming last: magnetism. At least a central experience from each topic is desired.

In attempting to shorten this syllabus, the teacher will have a hard time, as I have attempted to build up each topic carefully and thoroughly out of the simplest phenomena and thus establish a line of thought to foundational ideas. Thereby, a certain systematizing has entered in. Each proposition is prepared for and also proven from the preceding one. Such a structured form is indispensable for the teacher in order to introduce him to a new way of seeing; otherwise, he would have no solid ground under his feet for further questions by the students. He must be able to base his conceptions, carefully considered, on an extended field of perceptions. It's different for students: They certainly want to evaluate and establish, not out of a carefully controlled process of construction, but out of insights flashing up out of the circle of their own experiences. They certainly want hints and, upon occasion, an awareness that the concepts taught them are good judgments and are well-founded. But, they don't seek systematics or proof, rather a thinking which is experiential, primal and spontaneous—not a cornerstone for proof. Of course, proof must be given and the teacher must know it, but he should not lecture incessantly afterward.

Thereby it is clear how we can abbreviate: Dispose of systematics for the teacher; dispose of half the experiments. What appears to the teacher as most graspable, most characteristic, remains as a residue. That which is omitted can be recalled in great sweeping brush strokes and summarizing thoughts. For the few students who desire a definite systematic structure, or who want material for knotty reflection and love abstract summarizing-thinking, a great deal more of the systematically based material can be woven in on the side. With experiments, one should not bring in too much nor omit too much. For each experiment, there should be time to lovingly work it up, to present it in its fullness.

Grade 7 Overview

In contrast to the 6th-grade physics curriculum, in 7th grade we are more focused on the objective and apparatus-centered. We no longer consider just broad impressions or general experiential phenomena; now we focus on (1) more complicated apparatus manipulated by the teacher and the students and (2) their observations.

While the 6th-grade **acoustics** sequence developed out of music and then into pure tones, we begin 7th grade with pure tones but then focus on oscillation more closely as a material prerequisite of tone. We form mechanical concepts, yes, even abstract ones. Instead of clearly visible string lengths, we speak of invisible "frequencies."

In 7th-grade **optics**, we are no longer concerned with exploring the interweaving of the range of experiential brightness relationships about us, but now, using experimental lab apparatus, we generate new imaginations and concepts, initially using the mirror and the pinhole camera.

In our **heat** studies, we no longer just give a qualitative overview from decomposing heat stepwise toward form-building and rigidifying cold or ice. Instead, we measure the thermal expansion and thermal-insulation capacities of specific materials. Similarly in **electricity**: from a discussion of general electric effects such as the sparks of flashing electricity near streetcars, we shift to a study of electric current enclosed in simple circuits. And, we leave terrestrial **magnetism** aside to study the field of a powerful manufactured magnet and the way its man-made field behaves.

Grade 7 Acoustics

In grade 7, the various topics are arranged only minimally according to the 6th-grade sequence given by Dr. Steiner. Nevertheless, once again they form a series passing from the more connected, cohesive, qualitative-observational, into the more objective, manipulative, abstract and separated from the cosmos.

I. THE TUNING FORK

In this year's introduction to acoustics, we begin, not with a small string quartet, but with a collection of tuning forks on wooden sounding boxes (Ac1). The tuning fork was first invented by John Shore in 1711, around the time of England's ascendence. (See Section I under Grade 6 Electricity). How do tuning forks create their sounds? Where must they be struck, and where are they to be held firmly?

The way a tuning fork moves when sounding is demonstrated initially by striking it and then by holding it just touching the skin (Ac2) or touching the surface of water (Ac3). A base that resonates with the tuning fork augments its sound (Ac4). We work out how a tuning fork physically oscillates to embody sound, as the following illustration shows (deflections are exaggerated).

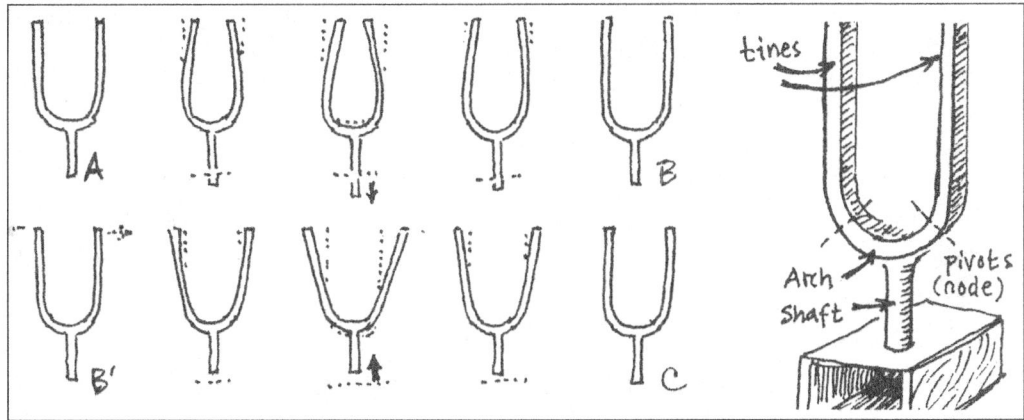

Figure 1. Mechanics of tuning fork oscillation

As the tines bend outward in the primary mode, the arch is forced open and shifts the shaft or stem upward. When the tines bend inward, it shifts downward. The oscillation of the tines (which occurs with a sufficient blow) is thereby transmitted through the shaft to a soundboard. This vertical oscillation of the shaft is damped only little by holding, so only a weak impact is needed.

The tines also oscillate in a second mode, moving transversely (one forward, one back); the deflection is greatest at the tine ends and is again null at the pivot point (where the connecting arch begins). At this node, the arch oscillates transverse to its length but with a small deflection. The shaft exhibits a longitudinal oscillation, parallel to its axis, although smaller than the primary oscillation of the tines.

The motion occurring from stage A to C (Figure 1) is termed one *oscillation*. Certainly, the initial position is regained at B, but the direction of movement at this time is not the same: At A, the tines are moving *in*ward and at B, *out*ward. The time interval corresponding to reach the same position and the same direction of movement is cycle time or *wavelength*. The number of oscillations per second is called *frequency*, expressed in units of Hertz (Hz).[1] 1Hz = one oscillation/second. Logically, frequency (cycles/sec) is the same as the reciprocal of cycle-time (seconds/cycle).

The so-called *overtones* arise if we strike the fork very hard. Superimposed on the basic tone (the "fundamental"), weak, shrill, higher tones are heard and quickly fade. The tines move in a complicated oscillation pattern with such overtones. However, if we strike the tine from the front (transversely), we produce mainly overtones (Ac2b, Figure 4b). Such a blow impacts the tines only briefly and thus is able to produce such very high-frequency overtones with their short cycle time.

With a moderately intense blow on the usual side, the tines produce a nearly overtone-free sound (not too great an amplitude). This "pure" tone is void, sterile and unchanging—unsuited for music and, therefore, more so for a complete study of acoustics.[2]

The pitch or height of the tone (i.e., frequency) of a tuning fork is clearly dependent on its size; or, more precisely, it depends on the mass of the tines and on their stiffness (the restorative force of the metal tines against bending). Long, heavy, flexible tines flex slowly to and fro, making a large excursion. In contrast, short, thick tines vibrate very quickly and with smaller deflection (Ac5). Therefore, if one attaches a weight (rider) as an extra mass on the tine-end, then the oscillations will be slower and the tone deeper (Ac5). Also, the extended contact with the rubber hammer in Ac3a & 3b acts to load the tine, favoring lower frequency tones.

II. OSCILLATION: TRACINGS AND FREQUENCY

1. OSCILLATION TRACING

A very pleasing curve arises when we draw a so-called "tracing" fork over a sooted glass plate (Ac6). It shows how the tines do not have a random zigzag, jerky motion, but come gently to rest and glide into moving the other way; all parts of the tracing are smoothly curved. The frequency (pitch) of the tines is independent of the amplitude (corresponding to the deflection). *Amplitude* gradually decreases while the frequency stays constant. We can deduce how fast our tracing fork was stroked across the glass by the number of peaks squeezed together in the tracing (i.e., the number of peaks-per-inch). The tracing on the left, i.e., in the diagram below, was made with a rapid stroke and that on the right with a slow stroke.

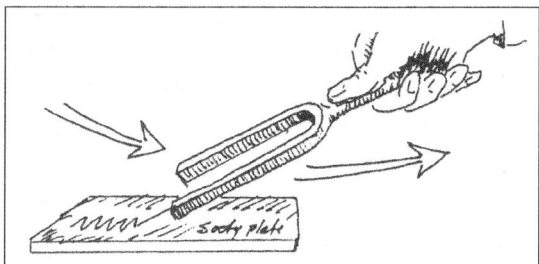
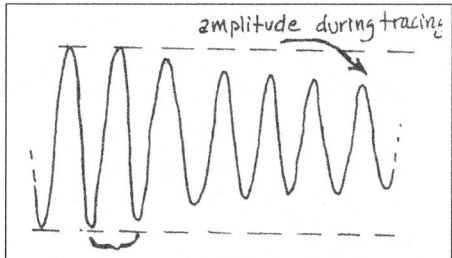

Figure 2a. Making oscillation traces   *Figure 2b. The oscillation tracing drawn with a slower stroke*

The width of the oscillation trace peak-to-peak reflects the excursion of the tines; the greater this displacement (*amplitude*), the louder the sound we hear. The displacement decreases more in the right tracing, because more time elapsed during this tracing—as is seen also from the greater number of oscillations (cycles) traced in the soot. We are attempting to make the details of the tuning fork's movement or oscillation visible in a new way with these tracings.

2. FREQUENCY MEASUREMENT

For measuring frequency, we must hold the *velocity* (speed) constant and measure it. It is simpler to rotate a disk with numbered teeth just fast enough so that an index card held against its spinning teeth makes a buzzing tone of the same pitch as the tuning fork. From the number of rotations and number of teeth, the frequency of the buzzing tone can be determined (Ac7). In principle, we could also cause the tuning fork to travel on a special runway at a constant, known velocity and use the tracing it makes under these controlled conditions to determine the frequency of a given pitch.

Acoustics

3. FREQUENCY OF TONES

The rate of oscillation of a tuning fork is too great to be visible to the naked eye. We see only that the tines are vibrating—they appear blurred—but we cannot follow the individual stages of the movement and thereby cannot distinguish the oscillations nor count them. Above 10Hz, all such oscillations are invisible to our eye. Only tones above 30 Hz have a distinguishable pitch to our ear.

Hearing and seeing are separated from each other by nature. Differentiation in hearing (high and low tones) cannot be directly replaced by differentiation in seeing. Nevertheless, if we imagine oscillations as visibly present where we actually cannot see them directly, then in thought we are no longer dealing with linking up sense-perceptions, but are hypothesizing something mechanical behind the sound. Although this line of thought is common in physics, we ought to be conscious of its problematic character. In teaching, this has the effect of conceiving a mechanical excursion or displacement as a numerical term for pitch, which we then make clear using mechanical apparatus and mechanical measurement. The way we measure frequency must be experiential at every step. Only then is the frequency concept clearly related to the revealed phenomena of sound, and the students do not learn a puzzling abstraction, for oscillation values of hundreds or thousands of Hertz are conceivable only in the abstract. Even more difficult to understand are the traveling compression fronts or spherically expanding pressure waves in the air. We defer a study of these to grade 8 when we study pressure in fluids.

4. FREQUENCY RELATIONSHIPS IN A MUSICAL SCALE

A look at frequency relationships corresponding to intervals of a scale can be a fitting conclusion to the 7th grade study of acoustics. The clearest (but more difficult) demonstration is to make dual tracings with two tuning forks stroked together and count the oscillations in each curve over the same course (distance); i.e., for 1 oscillation of the fundamental tone, there are 2 oscillations of the octave tone, and for each 2 of the fundamental, there are 3 of the fifth (Ac8). We can perhaps omit this experiment which requires specialized apparatus and simply read off the frequency stamped on each fork; so, when we hear the octave, the ratio is 1:2, for the fifth we see 2:3, etc. Or, we could recall the string-length relationships investigated last year and explain intervals in terms of a string half the length oscillating twice as fast, while one two-thirds shorter oscillates one-and-a-half times faster, etc. Expressed for frequencies, we get the following series:

G" A" B" C' D' E' F' G' A' B' C D E F G A B c d e f g a b c' d' e' f' g' a' b' c" d" e"
                ↑   ↑

    Octave (C to middle c): 2:1
    Fifth (C to G): 3:2
    Fourth (C to F): 4:3
    Third (C to E): 5:4.

This can be used to generate an 8-tone scale: C, D, E, F, G, A, B, C. This mathematically perfect scale is the natural or diatonic scale (see note under experiment Ac1, below).

The string length characterizes a relative interval step, while frequency designates a specific tone by a number. This is true whether we are dealing with air (in the flute), or with wire strings, or with a drum-hide. *[The life of music 'lives' within the intervals in this in-between, rather than at the quantifiable 'points' of the tones, with their measurable frequency.]*

In grade 7, we still ought to practice a qualitative listening as much as in grade 6. There it was easy, since we had music and the various intervals manifest visibly on different instruments. In grade 7, we often present the pure (even too pure and nearly empty) tones of the tuning fork, in contrast to last year's violin with its rich timbre. But even in the raspy tone of an index card against a toothed wheel, a trace of tone-quality can still be perceived and even be modified, depending on how one holds the paper. Such tones already show the transition to machines and motors where we have only a clattering, buzzing, or the like.

Experiments in Acoustics

AC 1 TUNING FORK CONCERT

Mount a set of commercial tuning forks on wooden sound boxes; strike the C E G, producing a chord. Lightly slide the mallet toward the end of one tine and the sound dies out, showing where the vibration is greatest—at the end. Now strike the fork again and note how the ringing fades yet remains unchanged (in timbre).

Natural tones, which form the diatonic major scale, (e.g., as with blowing on a pipe or flute or with overtones in plucked strings) cannot be maintained unaltered like this. To get such unaltering tones, we move into the realm of the equitempered scale, which is common today.

[Note: Often many lab (diatonic) tuning fork sets are not based on A = 440 Hz "concert pitch,"* giving C = 264 Hz rather upon C = 256 Hz (where C octave notes become power of 2). Then E = 320, G = 384, A = 426.7, and C' = 512 Hz. If your set has other frequencies, check the ratios between the tones. Diatonic ratios are: 9:8 and 10:9 for full tones and 16:15 for a halftone. The complete diatonic sequence is: C:D 9/8 (full), D:E 10/9 (full), E:F 16/15 (half), F:G 9/8 (full), G:A 10/9 (full), A:B 9/8 (full), B:C' 16/15 (half).

Alternatively, the forks in an equitempered set can also be based either on C = 256 Hz, but more often use A = 440 Hz as the basis, giving C = 261.6, E = 329.6, G = 392 and C' = 523.25 Hz. J.S. Bach proposed this adjusted scale so it would be easy to modulate to another key on a keyboard instrument without retuning it. Still, a master violinist will, for example, actually play D sharp differently than E flat, as is appropriate for the diatonic scale! [This information is presented more for the teacher than as a required topic for the students. – Transl.]

---

* Traditionally. But, in America, the trend is now to tune to A = 444 Hz, to produce a brighter and sharper orchestral sound. Europeans prefer a more mellow sound, and thus often tune to A = 430 Hz.

AC 2 MECHANICS OF RINGING

One purpose of this experiment is to carefully, thoughtfully consider observations in detail and lead the students to develop clear, phenomena-based concepts of the (invisible) mechanics of oscillation. The point is not so much that oscillation mechanics itself is such a major thing, but—in its simplicity—it gives us good practice in developing detailed, phenomena-based concepts.

a) A tuning fork placed on a sound box is struck with a rubber mallet from different directions; on the side at the end of the tine is the most effective place. This corresponds to the easiest direction of motion of the tines.

Figure 3. Striking tuning forks

b) If we strike the tine from the front instead of from the side, a high, shrill tone arises. It can only vibrate poorly in this direction and so buzzes or jingles.

c) The same result occurs if we reverse the mallet and strike the fork with the wooden handle. Note how the hard surface promotes high-frequency overtones: the first three—octave, 2nd octave, and octave-5th—are the most easily heard or recognized.

Experiments in Acoustics

AC 3 SPLASHING WATER

a) Using a freshly-struck tuning fork of not too low a pitch (e.g., G), just touch the surface of water in a bowl with the side of one tine. The water sprays out and makes a splashing sound (Figure 4a). As the tine is submerged, the pitch gets lower and fades away very quickly.

b) If we touch the front of both tines (Figure 4b), we see the water splashed to the sides and sprayed higher than before.

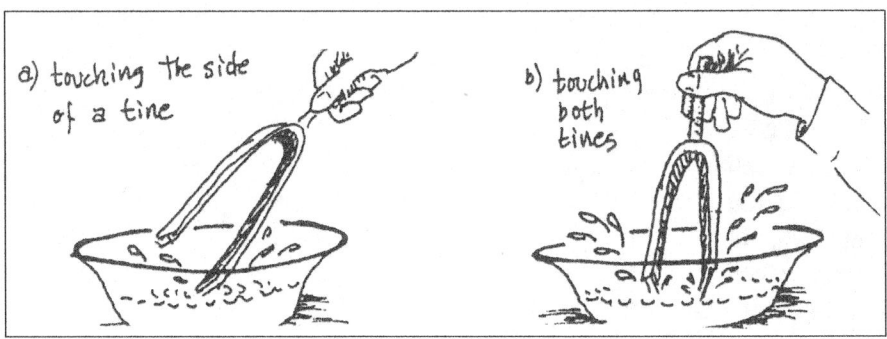

Figure 4. Showing direction of vibrations with water

c) If we hold a freshly-struck tuning fork as in 4a, but touching the tip of a student's nose, he feels an unpleasant prickling or irritating sensation. If he doesn't pull back, he will even get tears in his eyes from the irritation.

d) If we trace with a finger along the vibrating tine, we sense that the vibration is stronger toward the end and our touch extinguishes the sound most rapidly there.

AC 4 FUNCTION OF A SOUNDBOARD

a) Strike a tuning fork of medium pitch (440 Hz) and hold its shaft against a wooden soundboard (on a guitar or tuning fork sound box); the whole plate sounds. When the contact on the plate is poor (depending on the type of plate), the sound becomes buzzing and unclear. Retry this experiment touching doors, windows and railings. Note: With concrete we hardly hear any sound (the surface cannot resound).

b) A student listens first with his ear near a ringing tuning fork, then listens with the fork placed firmly on an intervening soundboard on his head. He will hear the tone clearly through his skull.

AC 5 HIGH AND DEEP

The length of the tines of the fork corresponds to the depth of tone. When we clamp a rider onto one of the tines, the pitch falls. The nearer the rider is to the tip, the lower the pitch. A very thick, short tuning fork of 2000 Hz (2 kilo Hertz = 2 kHz) must be struck with metal or hard wood, since the rapid oscillation must be initiated by an equally rapid blow. In water, such a fork does not splatter since the amplitude (motion displacement) is too small. Also, this fork only makes a soundboard resound weakly for the same reason (Ac4).

AC 6 OSCILLATION TRACES

A glass plate (24cm or 10" wide) is sooted using a candle. Take care not to hold the plate too deeply into the flame or in one spot for too long or the glass will shatter from heat shock; be patient, do a little at a time. Tracing tuning forks from commercial science supply houses are usually very long and thin, so that they produce a tone around 100 Hz, which unfortunately does not resound audibly for the class. One tine has a long needle which traces a beautiful oscillation line in the soot if the struck tuning fork is quickly and lightly stroked across the plate. (See also cello tracing in Ac8, below.)

[Note: Lacking a commercial tracing fork, try using the lowest in the lab set: C = 256 Hz, and tape a small needle onto the front of the tip. Note that the A fork (= 426.7 Hz) has too high a pitch and does not move with sufficient inertia to carry even the tiny additional weight of the needle without dampening. Also, try using a very small sooted glass plate (2" x 2") which will fit into a slide projector and project an enlarged image of the tracing onto the blackboard. – Transl.]

AC 7 FREQUENCY MEASUREMENT

For example, make a 20–30cm (8–10")-diameter disk out of pressboard, which has a soft wavy edge filed into the perimeter, consisting of 96 indentations (a multiple of 8 for easy purposes of layout and filing). Then remove the grinding wheel from a hand-driven grinding machine or hand drill and install the wheel. If we drive the wheel at a constant speed (e.g., 4 rpm) measured with a stopwatch, a small index card held against the edge will jump the number of times per second as bumps pass beneath it. Rotation times the count of bumps equals the frequency. An example formulated with fractions is:

$$\frac{\text{rotations}}{\text{second}} \times \frac{\text{bumps}}{\text{rotation}} = \frac{\text{oscillations}}{\text{second}} \quad \text{or} \quad \text{Frequency (Hz)}$$

The index card makes a buzzing sound approximating a tone corresponding to the frequency of bumps (e.g., diatonic G = 384). Compare the pitch of various tuning forks and see to which the card comes closest. For example, measure the rotation more accurately by marking the time of a group of 10 rotations, then divide by 10 to get the average revolutions/second.

AC 8 INTERVALS & FREQUENCY (OPTIONAL)

If anyone wants to attempt a tricky but graphic experiment, there are various possibilities. They can saw and file off just enough excess length from the tips of an extra 100 Hz tracing tuning fork to produce one with a frequency of 200 Hz, then attach the tracing point with glue. When stroked together with a 100 Hz fork over a sooted glass plate, the ratio of oscillation peaks per inch (2:1) shows very clearly. It is also impressive to glue a needle tip onto a cello string, stroke it gently with a small glass plate (while bowing the string). The closely-packed oscillations can be projected to a large size for the class to see.

[Note: More work on resonant tuning forks can be done in 8th grade acoustics.]

Grade 7 Optics

INTRODUCTION

In the 6th grade, we experienced the weaving of brightness in between the heavens and the earth, between sun-brightness and the human eye. What we observed was connected through thinking, linked up and ordered into a series. Also, the gesture of shadows was brought as one example of the way the sun and its shadows are continually transforming one image (the objects) into another (the shadows).

In grade seven, too, we should deal with light-images, not with light rays. Such rays can be sketched out all too quickly. To organize images into a coherent series through thinking demands that one advance beyond the details and grasp the concept already in what we are seeing, rather than obtaining it as an afterthought using abstract ray-constructions. This requires more time and thus the optics in 7th grade claims almost 20% of the block (at least one week).

The image-relationships of 7th grade are quite different from those of the 6th. The images we studied then, the so-called scenes we viewed, were seen as a whole (e.g., point-source lamp and shadows)—we were inside the scene. Now in the 7th grade, we will work out the process of scene formation from another approach: by means of a "symmetry-principle" (mirror) or reversal-laws (pinhole camera). We are now more outside observers.

Now, thinking not only passively follows the observed phenomena, but inwardly works with and sculpts them. The phenomena of fuzzy shadows and shadow-images, which came to the fore in 6th grade, will—in the 7th grade—be carried into the first clear pictures where color and spatial experience are cultivated. The artistic creation of images is the chief domain of optics today and is considered its task.

But comprehending and living into the images created by nature or by the working of the sun is of no interest to people other than "peasants and sailors." Yet, this will actually be our chief task. However, we must make compromise with the students. In lieu of the world's picture-creating, we will study the manipulation of images made by people using their intellect and will. The pinhole camera can also be used as the introduction to photographic apparatus. Because the mirror is less glamorous, it probably should come first.

Optics

I. MIRROR PHENOMENA

Human cognition and the way we view the world often models the world as a creation of mirrored-images; the human being mirrors the world in his thought-images. In the face of such casual popular ideas, first of all we ought to clarify the natural phenomena of a mirror, the conditions under which a mirror image arises, and how they are experienced. There are many preconceptions here; the mirror is not merely another topic in optics, but it also crops up frequently in mythology and fairy tales. Therefore, what follows is a far-ranging consideration including myths and poems about mirrors. Still, a complete study of this would go beyond what could be actually presented in grade 7. Rather, we seek mainly to open up an experiential gateway into optics.

1. EXPERIENCES IN THE VICINITY OF A MIRROR

Is the mirror a tool merely to redirect our gaze? Does it cast the impinging image back unaltered? Or does it create a characteristic image of the world, which would not arise without it? The simplest mirror is created by the earth itself where it is covered by a water surface. If the play of elements subsides, if no waves run toward shore and the sea is not disturbed by wind, then, in a totally unchanging stillness of water and air (Op1), the width of the heavens appears, but seen deep under the surface of the water! We look among the clouds "down" into the blue of the sky in which the sun shines. A second, "yonder world" seems to open up—as in the fairy tale of Mother Holle, where the girl falls into another world through the reflection on the water in a well. In this "pond-reflection," the living, green earth seems to swim at the interface of two heavens; its heaviness seems to be dissolved. Also, the perceptive person is caught up in this experience of weightless space, leaves behind his sense of being closed up inside an earthly body. The scene seen below becomes a pure image of the heavens, which we otherwise have above us—or are accustomed to have above us.

If we really experience such a mirror-stillness of the sea, then we might feel threatened. The ordinary life of nature's elements seems to have died. Also, as the water's surface becomes still and dead, it disappears.

> Deepest stillness reigns o'er water,   Not a breeze from any side!
> Not a movement stirs the sea.   Death-like stillness, frightful fell;
> And the sailor with distress   In the vast expanse of oceans wide
> Sees it glisten round him there!   Comes no wave to break the spell.
> – Goethe, "Ocean's Stillness"

In a short essay[1] Aleksandr Solzhenitsyn describes from two sides how the God-fearing person, submitting to the divine, can, like still water, be a selfless mirror of the highest. And, being still, does he not also show a divine constancy?

Why is it just water that is able to be a mirror and pull down the heavens into an image in the depths? With the blowing of the wind, it disappears (evaporates); soon it reappears and trickles into the earth, flows together and forms an ephemeral surface beneath the heavens: so uninterrupted above and connected below—like its mirroring surface is for our eye.

2. THE HUMAN BEING BEFORE THE MIRROR

In one sense every mirror surface is something disembodied, unearthly: It presents a totally undifferentiated surface. It exhibits no directionality in its smoothness. With water, it presents no resistance to slow movement in any direction. With metal mirrors, their glassy surface offers no impediment to gentle stroking. In attaining a polished state, glass loses its usual earthy roughness. Each stroke of the polishing wheel as it rubs over the still, matte surface grinds every large grain into powder as fine as dust and removes the tiniest ridges and valleys. There arises a glistening plane surface as if on a giant crystal. It must be as ideal as possible or else it will exhibit distortions in its image, which will be very noticeable in detailed scenes. Well-polished, large glass panes lead to the term "having a mirror polish." Thus the plane mirror points out of the domain of uneven surfaces that alter the image, to the realm of the crystalline, of ideal surfaces of minerals and precious gems.[2]

So-called float-glass, which has been manufactured by pouring soft glass out over the surface of molten tin—and today rivals polished glass in quality—actually originates from a piece of the earth (the tin surface) which has been brought to the state of a total mirror of its environment (by melting to a fluid state). The tin surface is a far more perfect mirror than water, and thus is almost invisible.

Often, when plant leaves create a mirror effect, we call them shiny. Their otherwise green surface is taken over by a whitish brightness—which however, vanishes if the viewer does not look again from the same viewpoint. Therefore we call this "directional-brightness" (see my discussion under Grade Six Optics, Section 1.5). But this directional brightness in plants never forms a real mirror image. Only wet, water-covered surfaces, e.g., streets in rain, show a transition from directional-brightness to mirroring. Ordinary surfaces can also act as mirrors if we look obliquely (nearly parallel to the surface). Most dinner plates viewed thus will show a lamp mirrored in it as two lamps. With our eye looking obliquely (and from

Optics

far away) these two almost fuse together into one symmetric image: The glazed plate approaches a mirror.[3]

An interesting demonstration of another view of a mirror is made by slowly clearing away talcum powder strewn over a mirror (Op1b). The dusty mirror is side-lit, with a lamp above and shielded from the darkened classroom. The powder spread on the mirror is now "co-bright" (see discussion in grade 6, Section 1.3). Now, as we remove the dust stroke by stroke, the "polished" surface becomes dark, in contrast to the surrounding areas (i.e., the ceiling), which appear more brightly illuminated! From our location we can see a lamp in the mirror; or, more exactly, we see it as if framed in a "window" (the mirror edge), which looks into a "room beyond the mirror." The mirror surface is now invisible due to its smoothness, and we usually can see beyond it into the entire scene, to the dark wall of the classroom, just as if we were looking through a window. Something tangible (the mirror) becomes invisible; something visible (the scene in the mirror) becomes intangible! That which we see in the mirror as through a window, however, cannot be reached as if it were a door. Still, it has a kind of reality: The lamp seen through the mirror-"window" (mirror-lamp) actually illuminates my ceiling and makes the still dusty areas of the mirror-glass surface bright from beneath, just as if it were shining up through an ordinary window (the mirror frame).

It is even more paradoxical if we place a hand over the mirror. Two shadows now appear, one cast from my hand by the mirror lamp onto my ceiling, another from the hand visible in the mirror by my lamp onto the mirror ceiling! (Really spend time viewing and pondering this scene.)

Thus, visible- and tangible-space (space in which we can move) are sundered. To the naïve understanding, such a situation is certainly confusing, even dangerous. Illusory images appear, "windows" into unattainable spaces open up; science fiction authors use such mirrors to call up images of distant planets, growing large as if approached, which then pop out of the mirror to this side.

This uncoupling of visible from tangible space (see prisms, Grade 8) is not dangerous with a water surface. Since the heavens are intangible anyway, what we see in the mirror image wouldn't be taken for something tangible. Seeing it below us lets us quickly recognize this new situation, because we are accustomed to a world organized from above downward. So, a horizontal mirror carries us off, floating, yet noticeably. And, in contrast, the vertical mirror-"window" deceives us. For, as soon as I see the usual earthly objects in a vertical mirror, I can think that they are where they are seen through the mirror-"window," when actually (tangibly) they stand somewhere else.

The horizontal mirror (the "lake mirror") pulls us into another world. The fairy tale of Mother Holle can develop a rich image of this for us as intellectually oriented observers, yet weak in experience. There, we have the image of the reflection in the well as the portal to Mother Holle's land. In contrast, a mirror on the wall can insidiously and maliciously replace the otherwise reliable world, falsely switching the place of things, mirroring outer, illusory places. And, when I look at myself in such a wall mirror (perhaps with a sense of enjoyment), then it is possible to experience myself as having just become my physical countenance. My inner soul-forces and life's course have made an imprint on my countenance. It is another matter if I wish to understand myself from within, rather than outwardly as in a mirror. Within I become more myself, since, in inwardly conceived soul-activity, I gently, delicately transform myself into the future (striving toward an inner imagination of my future).

Thus, the outer mirror image directs itself counter to these future-forces and binds me to images of the past. Once again, this is hinted at in fairy tales: "Mirror, mirror, on the wall" calls the witch in hatred against the maturing Snow White, against all that which leads to coming generations. Also, it is a mirror which proffers Faust[4] a glimpse of woman (Helen) as an idealized, ancient and unchanging vision. Or, recall the picture of Louis XIV, the Sun-King, in his hall of mirrors (which at that time of hand-polished glass, was truly a marvel), saying "I am the state." He surrounded himself with ministers and generals in courtly attire, reduced mirror images of his eternal nobility. We see how the sundering of visible- and tangible-senses draws us into trouble.

3. WHO KNOWS THE MIRROR IMAGE?

Place a large vertical mirror in the room in front of us (Op2, the second, or window situation). We can see part of the classroom in it, as if through a window. That which we can see in the mirror, we could also see behind us if we turned around. Initially, we don't turn around, don't move back and forth, but simply view what we can see in the space beyond the "mirror window." The furniture, the door frame, and everything in "that" room corresponds to what we know is in our room. They each have a brighter side toward the direction of the opening outside (the windows), though perhaps invisible to us. All of the space is illuminated—in the same way we know this actual room in which we move about is illuminated. The visible space yonder and that on this side of the "window" (the mirror surface), both appear to be the same with respect to the lighting. And, yet, correspondence is not complete.

Optics

A mirror set up outside in a field would show us the landscape (the one lying behind us). But that scene appears abruptly at the edge of the mirror without transition from the surrounding landscape (visible in front). At the mirror's edge is a break in the unity of the scene in front of us; this we could interpret as a kind of "window frame." But this "window" is really strange: It apparently hangs in space with no wall to support it! If I made a wall for this "window" (i.e., built one around it), then I only formalize this break since, when we approach an ordinary window in a wall, we expect such a break at its frame, usually opening onto an outside landscape. When I turn around, I also see a landscape: This "window" does not open from inside out to free nature outside, but outside to outside! I look out (via the mirror), yet I stand outside! Moreover if I slid the wall around behind me, the landscape seen in the "window" disappears and another room appears (I'll see the wall). The mirror forms a "window" between two rooms. Normal windows stand between an inside- and an outside-space; but, it is just this which is impossible with a mirror-"window." The mirror cannot be used as a true window, joining spaces distinguishable as experientially inside and outside. The situation created by the mirror has few counterparts in nature: One imagines a gateway in the mountains where one canyon opens into another, or a ventilation trap in a mining tunnel or a ticket window between two offices. So, approaching a mirror, we notice (via the context surroundings) that the segment of the world we see is not seen through a window or glass porthole; i.e., this is no ordinary window between inside and outside. Rather, it is a more profound displacement than is created through a usual window.

Up to now, we have spoken of a mirror-"window" as if the usually established flat mirror image did not exist. That is correct: What we see in a mirror is never a planar image, lying somewhere on the surface of the mirror. Rather, a space is opened up, which—for vision—stretches beyond the mirror. The accommodation of our eye's lenses and the binocular alignment of our two eyes is carried out in the normal manner as we peer into this (apparent or virtual) space "behind" the mirror. We can perform the most diverse experiments in connection with this: This "yonder space" (beyond the mirror) possesses not only spatial depth, but is visually equivalent to a completely normal space. Not only the illumination and spatial depth of things seen yonder are equivalent to the relationships on this side (as we saw), but also the way near and distant things obstruct one another in the space yonder when I move my head side to side corresponds to the way it occurs here. Thus, each yonder object shifts in accordance with its apparent distance in the space beyond the mirror (parallax). When I move side to side, things which stand close to the "window" (the mirror) and thus appear close to me, shift a lot (relative to the "window" frame), while distant objects are shifted less. Even the edge of the mirror is included in this "correct" visual depth. If I shift my head to the left, then a new portion of the yonder space appears on the right. This differential shifting of close and distant visual objects in connection with a changing viewpoint should be noticed every time we take a walk—all consideration of the mirror aside. Then a person will discover in the mirror image that it truly presents a mirror-room to our eyes.

4. THE BOX YONDER

Anyone who studies the mirror will ask: "How are mirrored objects related to those in front, to actual objects?" This question cannot be answered in this form, since—the commonplace conception that we "see the same thing, just in a mirror," notwithstanding—we must first attain a consciousness of what we actually see in the mirror. With this search, we must lay aside the everyday term "mirror image." Instead we will speak of mirror space, which we call *yonder space*. It is optically (visually) complete, though fundamentally unattainable, on the yonder side.

If several people sit in front of a mirror, then they see a scene peopled correspondingly with people. The relationship which obtains between these two sets of people, whether similar or corresponding, is pushed into the background for now, and we ask a much simpler question: Do all those sitting here see the

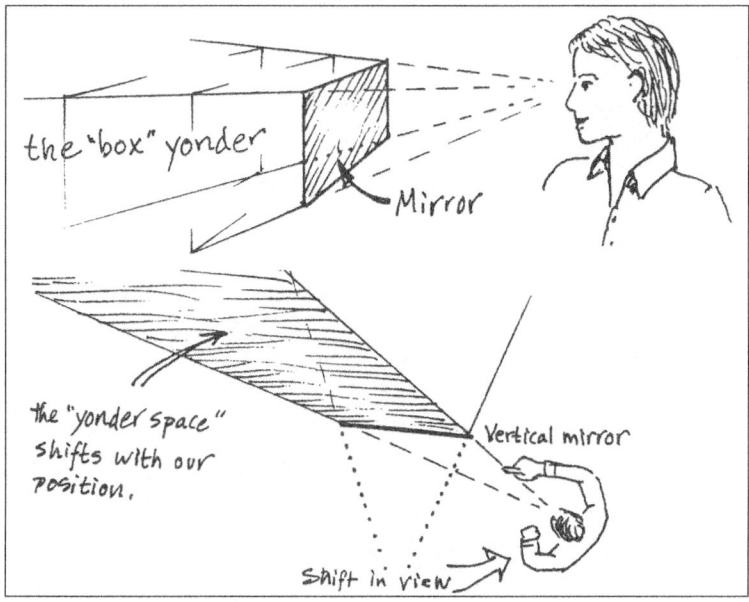

Figure 5. The box yonder

Optics

same thing? No, they do not, since the mirror edge cuts off a different view into the yonder space for each one. In the middle of the yonder field of view, there is a broad zone that is seen by all; at the edge, it is different. This can be imagined in a tangible way as the objects visible yonder being located in an individual box which begins at the mirror and stretches off to infinity. The edges of the mirror form planes determined by the position of the mirror and the individual viewer (planes determined by a point and a line). A third observer could project the yonder box of a viewer in the following way:

If the mirror has no frame, the box yonder begins in the space on this side where the edge of the mirror obscures the view of this side-space. In between there is a dead zone; the yonder space replaces abruptly this side space. If the viewer shifts sideways, the yonder box shifts laterally also, shoves itself in front of things formerly seen on this side and, with a diagonal view, becomes contracted and smaller. It shrinks to nothing when we move behind the mirror, behind the extended plane of its surface. Therefore, no one can enter his own yonder side (since it is only our own, individual box yonder which we can see). The closer we stand to the mirror, the more the box yonder spreads out into the distance as a truncated pyramid, finally including 180° when we are at the mirror. The box yonder cannot include more than this for a plane mirror.

5. THE BOUNDARY VANISHES

There is a simple but surprising experiment (Op2b) which is demonstrated as follows: Place a tall, burning candle close in front of a mirror on a tabletop covered with a white tablecloth or paper. Somewhat closer to the mirror and away from the candle, place a figurine (nutcracker or Christmas figurine, etc.). The shadows created in the yonder space and in this side space pass over the boundary between this side and yonder side without breaking or bending!

Figure 6. Shadows across the boundary

It is as if the mirror has vanished and we have in front of us a table (half on this side, and half on the yonder side) with two candles (one this side, one yonder) and two figurines. The shadow is deeper at the spot (3) where the shadow cast by the candle and the figurine on this side overlaps the shadow cast from the yonder side candle and the yonder side figurine! This doubly-shadowed spot (3) always lies symmetrically on the border. We know how the shadow law[5] applies on this side: A surface is darker than adjacent ones (more shadowed) if, looking out into its cupola, one can see less brightness than from adjacent places. The spot is darker because from there (spot 3) the candle cannot be seen.

The proof of this law on the yonder side is effectively shown by "looking" from the yonder side. Though this seems ridiculous at first, a view out of the yonder side can easily be arranged by utilizing an effect which we will later have to investigate: Something will occur in the yonder space if it is done on this side. So, we send an experimenter into the space on this side and ask him to move his head slowly about the table surface, letting us know when the figurine on this side masks the candle on yonder side. Then, see if his face is in shadow at this spot. Simultaneously, a matching experimenter appears in the yonder space; we see him speak and see if his face is in shadow also—the correspondence between this side and yonder side holds exactly. Initially, we could suggest that the report of the yonder experimenter that "the figurine on this side conceals the yonder-side candle for me" is wrongly expressed. But our objection rests only on the fact that we have termed our standpoint "this side."

The yonder observer reports correctly from his viewpoint, since over "yonder" is "this side" for him, and to him we are in the yonder side. From a purely visual standpoint, there is no significance in viewpoint: "this side" and "yonder side" have equal validity. For our experimental situation, we can really speak only about space A and space B. We feel intuitively that space A is self-evidently different from B, that one is actual and the other is only apparent. But, aside from this feeling, we inevitably will have to explore how we do know anything about the world (epistemology).

A doubter might now object: 1) Isn't our "this side" recognizable by the fact that we can move about in it? We have already seen how we can visually move about in the yonder side and how optical laws hold true there as well (the overlapping of shadows). 2) Doesn't the mirror plane form a boundary impenetrable to all movement on this side, which thus is distinct; i.e., we can only pound on this side of the mirror, it seems? There is no movement that we can see only here on this side: Someone on the yonder side also pounds on the mirror. It is even impossible to base the distinction on my vision, since for pounding, I feel my fist collide not with a surface, but with the fist pounding from the yonder side! Even the smooth feel of the mirror is perhaps due only to the constant contact of a corresponding finger yonder!

Also, the feeling of coldness on touching the mirror surface lets the mirror film approach the coolness of air. In any case, the feeling of warmth of the yonder fmger is not perceptible. The yonder finger is felt as if through a thin film-like air, or like nothing, but without body warmth. We can summarize this: There is one invisible boundary in movement between the two spaces; however, movement on both sides corresponds exactly, for vision it is exact.

Some may still be troubled by the questions as to how the spatial distances yonder compare with those on this side, whether the objects appear undistorted and the same size. To verify this we should place a yonder object next to one from this side; but that can never happen! Something else makes it simpler, however. We lay a yardstick on the mirror and simultaneously there is one on the yonder side. We can see both and their markings correspond exactly. One could object that distances correspond only in two dimensions, vertically and laterally, but not in spatial depth. The answer to this can best be learned from viewing without bias: We can directly see the exact correspondence of depth with only a little practice. However, we can also confirm our impression by the following demonstration: Place a vertical pole (classroom flag stand) in front of the mirror. A similar one now appears in the yonder space. Then place another vertical pole somewhere behind (beyond) the mirror—a "sighting-pole" tall enough so that it shows above the mirror's top edge—and placed so it is seen in the yonder side box. Move it closer and further away, until the top poking above the mirror now shifts exactly with the reflected pole, as it shifts in the yonder box, as we move to and fro. We can now measure the distance to the pole behind the mirror and find that it will be exactly equal to the distance to the pole in front of the mirror. The distance to the base of objects on the yonder side of the mirror is equal to the distance to the corresponding object on this side. The mirror offers another world of visible things, situated in a mirror space. But one cannot really see this if we constantly think of the mirror as a tangible surface.

6. ONLY THE BLOOD RESCUES US

If we focus purely on seeing and take it seriously, then with careful consideration of the relationships, we fall into confusion. Conrad Ferdinand Meyer affirmed this in a poem—and also hinted at a way to overcome it:

Flight of the Gull (1882 edition)

    A gull I saw 'round rocky promontory circle
    Untiring fly the same track round & round,
    Soaring full of life on outstretched wings;
    And all at once, below, in sea-green mirrored
    The selfsame craggy point I saw encircled
    With shining wings outstretched in fleeing
    Which flashed untiring through the water's deeps.
    And the mirror made for me so clear an image
    That no distinction twixt the wings below
    And those up high and free in air was given;
    And Illusion and Truth became the same.
    Gradually was I overcome with dread,
    The Essence and Appearance so akin to see,
    And tarrying on the strand, I asked—
    While staring at this ghostly scene—
    And you, yourself? So, are you truly winged?
    Or merely painted like a mirror image?
    Do you juggle in a circlet, fairy tales, and legends?
    Or truly have you blood within your wings?

The poem ends with a question. How can I maintain my self in my identity in the face of a mirror image?[6] We can well consider the importance of our warm blood and of its forces, its power and will. In an earlier version (1881) of this poem, the last refrain was "Or have you power within your wings?" (instead of blood). I must have reached a point of certainty that, when I see myself here and in a mirror image, it is I who inwardly animates this body filled with warmth and blood. And, when I hold my hand before the mirror and another hand comes toward me from the yonder space, moving exactly the same as my hand does, then I can safely ask myself, "Have I moved two hands or one? Which of the two is my body, and which is some other yonder person's, in which I do not live?" Visually it is not too clear, but to our blood-filled limbs it is.

How does my resolution connect with my body? It must be understood as connected with a self-conscious being, incarnate in a blood-filled body, with senses for balance and self-movement. Then the illusory qualities of the mirror collapse, and we can fundamentally distinguish this side from yonder side—though only by

Optics

setting aside our visual sense. Since we are now able to distinguish a clear difference between this side and yonder side, with the help of our lower senses (balance and self-movement), it makes sense to investigate the relationship between these two sides in connection with other things we see, i.e., to inquire into the structure of the mirror image.

II. THE STRUCTURE OF THE IMAGE WITH MIRRORS

1. INITIAL FORM OF THE MIRROR LAW

Up to now we have spoken constantly about the similarity of the yonder space (seen in the mirror) and this space (seen directly). But, are they totally equivalent? Do we see there the same things as here? No, that is not exactly the case: First of all, the yonder space is always seen from a different viewpoint, and secondly, it is seen right-left reversed. Although there is not complete equivalence of phenomena, still a precise law connecting both still holds. How can we express this? We must first ask: From what position should we view an object (on this side), so that it would appear the way its counterpart is now seen in the yonder space.

To answer this, we can do the following experiment (Opla), hopefully an everyday experience. We shove our fingers up over the edge of an unframed horizontal mirror:

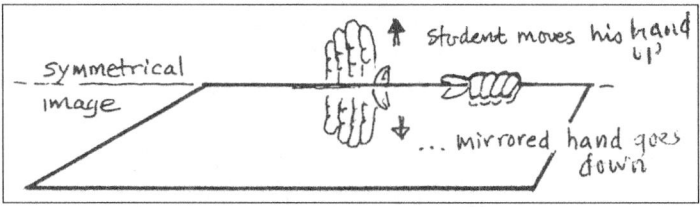

Figure 7. Pond reflection

Looking obliquely (nearly parallel to the surface), the clump of fingers on this side and yonder side appear identical, since they are seen from (essentially) the same direction. A symmetrical scene appears, more striking with a larger mirror. Any visible edges (such as the sides of the ingers) which run vertically into the mirror edge, then cross over seamlessly into the mirror image, and the boundary of the yonder side becomes invisible! For things near a mirror edge, one sees a completely symmetric scene. (This is true not only for silver-plated or front-surfaced mirrors, but at this oblique angle, even the usual rear-silvered mirror will reflect primarily from the upper glass surface.) This is even a universal natural law: Viewed obliquely, water and every wet surface become almost perfect mirrors. For a squatting observer, the surface of a pond nearby is somewhat transparent, but in the distance reflects metallically. Crouching low, the colors in the mirror scene out there are just as bright as the original.

Looking from above the mirror, however, the observer does not see the fingers yonder directly but sees it from another direction than the fingers on this side. No matter how we look at the mirror, only a symmetrical situation of things seen this side and yonder holds true, but not a symmetry of point of view. Simply put, we constantly see a yonder side twin standing directly opposite the corresponding subject on this side, though the twin is seen from a different point of view. From observing many such symmetries, we recognize the ideal symmetry-plane: the mirror surface. For vision alone (especially with large mirrors) the surface is only discernible by means of this symmetry. A line (in a tangible space) joining the corresponding pairs crosses the mirror at right angles and is bisected, this side and yonder. In order to see someone standing next to me on this side the way I see them in the mirror, I must look from the corresponding location of my twin; i.e., perpendicularly opposite the original and equidistant to the mirror surface. Or, I can imagine the scene yonder in the mirror as made up by replicating objects perpendicularly opposite and equidistant onto the other side of the mirror (aside from the right-left interchange). However, instead of imagining objects being replicated over yonder, we can leave our place on this side and shift the observer into the hinterland (formerly concealed by the yonder space of the mirror) where he will see things on this side the way we saw them in the mirror.

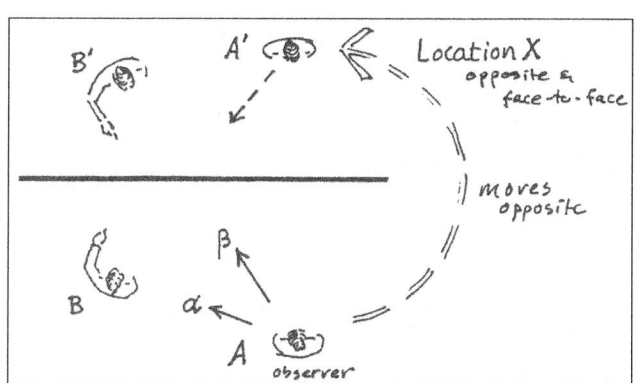

Figure 8. Ski view of alternate observer positions

Observer A stands in front of the mirror next to person B. A sees B either next to him on this side, with line-of-sight (in profile, looking at his ear), or as B' (mirror-twin) yonder, with line-of-sight (looking him in the eye). In order to see B from the same viewpoint as we saw his twin, we must go over to a place X perpendicularly opposite our initial place. Does such a rule for the position of things in a mirror express an archetypal phenomenon of mirrors?

Optics

2. RIGHT-LEFT REVERSAL

The mirror law we have worked out so far is unfortunately not purely optical, i.e., is not clarified in what we see. It is based much more on correspondence in measurements (distance opposite) than on translocation of objects. Moreover, it is only half the story, because everything gets mirrored to us right-left reversed! A yonder subject, i.e., B' (the twin friend in Figure 8) would be extending his left hand, as our friend next to us extends his right! And, there is no position from which we can view B so that his extended right hand becomes his left—as it was with B' yonder.

If we cannot avoid the phenomena of right-left reversal, we might perhaps attempt to push it under the rug by rationalizing, in order to salvage the mirror law worked out above in elementary form. "The person B' seen yonder lacks not one freckle, not one speck is added or omitted which can be seen on B on this side. We could consider the right-left reversal as something negligible, as an added complication to be overlooked." So saying, we would neglect not an incidental aspect but a constituent part of mirror phenomena. Instead, it would be valuable to examine what the reversal signifies for the appearance of objects.

It is well-known how a right glove can be transformed into a left only by turning it inside out, though the lining now is outside. But a flesh and blood hand cannot be turned inside out. If, nevertheless, we wanted to transform or "fold inside out" a model of a person, then we would have to place him in life size next to a plane and replicate each point on his body perpendicularly opposite and equidistant onto the other side.

But the left shoulder point will become the right shoulder point. The whole body must be inwardly dismembered: Everything must penetrate through everything else. This is completely impossible, at least with any natural form. The first way we can do this is with raw

Figure 9. Perpendicularity opposite duplication

egg white, which "slides through" so that the rear is dragged to the front. In two dimensions, it is usually enough to lay one form over another, flipping planes far apart on top of one another: The right & left will always reverse since the flipping-over corresponds to mirroring: The form is brought perpendicularly opposite. But, for actual bodies of flesh and blood, it is impossible.

Everyone soon runs into the question: Why does the mirror switch right & left and not also switch top & bottom? A little consideration is needed here as to the forms in nature which have a right & left. A sea cucumber, for example, which only has a mouth at one end and otherwise is like a barrel (has rotational symmetry), can never be described as having been injured on the "right side," for it possesses neither right nor left. If we grasp it differently in our hand, the "right" side would disappear. Only when an additional front/back element (e.g., a dorsal fin) is added to up & down is one side permanently recognizable. Only when two different sides of the body are present can a third be determined by means of them. With a form that differs front to back as well as top to bottom, we are able to distinguish a left or right body side from both these poles. This distinction occurs not directly by virtue of an outer variation between right & left body halves; rather, we prefer to speak of right & left, not when there are outer distinguishing features, but more when we realize an inner organization in relation to other directions already endowed with outer distinction.

How is something inwardly organized? A right-hand screw, advancing frontward, will point "up" after a quarter turn "right" (i.e., clockwise); a left-hand screw, in contrast, will move frontward ("advance") when turned "left." The archetypal form of a distinction in three dimensions is a right-hand helix and left-hand helix! There are no other kinds of helixes. The distinction left-right also turns out to be based on our sense of rotation; only through this sense are we inwardly sure of our directions when we establish right-left out of a given arrangement of front & back and up & down. If I put myself in the position of someone standing obliquely to me, in order to verify whether the arm I see is his right one, I use my movement-sense and a right-hand spiral motion, e.g., with a tiny hand movement.

Summarized: Right-left is not an outwardly determinable direction axis, like one of the three equivalent Euclidean axes in space but, rather, marks a distinction within the body, independent of its orientation to outer directions in space. An inner helical relation of axes can be described only in the actual body. In this we always mean: helix, rotation, movement. This involves an adjunct movement, which occurs over time. Thus, right-left distinctions do not point to outer spatial configurations, but point to something temporal.

Before we reach a conclusion about this question, we should first further explore the relationships of space. To utilize a distinction of right & left, the body which we want to describe with right-left must carry distinguishing features in two mutually perpendicular directions. An egg has no right or left. If another direction can be distinguished on a body, after top-bottom, the next one is named front-back

Optics

(cf. a seashell). If a three-dimensional coordinate cross is labeled with "bottom" and "front," then the third arrow will unambiguously point out a "rightward" direction. (Figure 10). If I switch only one of the other labels, then right-left must also be switched; if I switch both simultaneously, "right" would be retained.

Earlier, we asked the question: Why does the mirror reverse right & left and not top & bottom, since it stretches like a plane upward and to the side in space? For the first time, we can now form the answer. Outer spatial directions which lie in the mirror-plane are not switched at all! If I look exactly north into a mirror, then I see all that is east of me also east of my mirror twin; the same for things above or below. Everything seen in the mirror space lies in the same direction as in the space on this side.

Now, we have to ask, how can the mirror switch anything else? [We define all directions relative to the viewer.]. If I stand in front of a mirror wall, extending from east to west (Figure 10) and climb up a stepladder, my mirror twin does the same. But, if I step northward (toward the mirror), the twin doesn't step north (in the same outer direction); rather than retreating, he comes toward me, to the south:

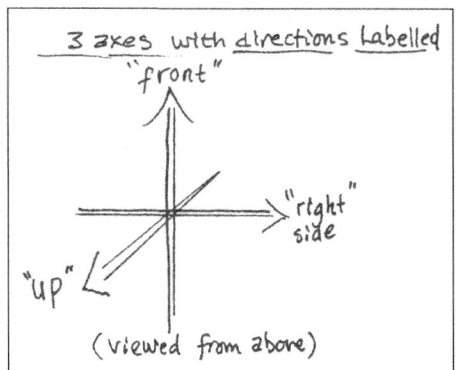

Figure 10. Directions in mirror-space

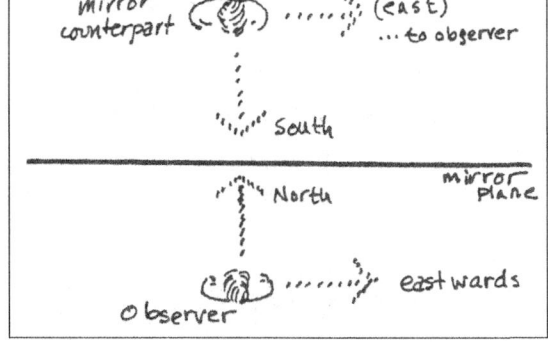

Figure 11. Directions about observer

Thus, one outer direction of space is reversed; this movement done by the yonder person is now performed reversed. But we don't need his movement to see this; it is already visible in his orientation: His chest side is toward me and his back to (my) north. One direction relative to the body itself is twisted about: the coordinate cross above (Figure 11) can be used to show how north & south (front & back) changed places.

Step-by-step: (1) I imagine a figure of three axes in space. (2) I can attach a direction-label to each line. (3) A right-hand thread screws "forward" when I rotate it one-quarter turn from "above" to the "right." From this we have: Sideward means rightward (with respect to above and forward). (4) Now I switch "front" and "back."[7] I use the same right-hand screw [held the same way] and screw it "forward" now, i.e., toward the *new* "front"—toward what *was* "back." Now I must turn the screw from above to the other side. This "other side" must (by our earlier definition) be "right" since it is a right-hand thread. However, if I screw the thread so it advances toward a mirror, the direction of advance in the mirror space is effectively reversed: I see a mirror screw, also rotating rightward, advance toward me—so, it must be a left-handed screw. To get this screw to advance "frontward" (with respect to me), I would (from my viewpoint) have to turn it a quarter turn from "above" to "left" (5) If I now switch the labels "below" & "above," then sideward can no longer be labeled "right," but must be "left," or else a right-hand thread, advancing toward the mirror (now "back") would be rotated to the inside we still had labeled as "right." Whew!

The solution to the problem of mirror-reversal—once all this has been understood—lies right before our eyes (no pun intended). I move forward in front of a mirror, and the yonder person comes toward me. In this everything is contained: one outer direction is reversed, and thereby an inner direction of the body is reversed, and the screw becomes the opposite. I can hold a screw in any desired orientation before the mirror, and I will see the opposite thread in the mirror space. This is the Goethean method: to acquire knowledge not via apparatus, but primarily in a wakeful experience/observation of what actually happens.

3. A PURELY VISUAL MIRROR-LAW

After we have become familiar with the right-left reversal in all its features and have understood how it is an inseparable part of the pictures that occur with mirroring, it is worthwhile to express this in a mirror-law. *From what location should A look toward B on this side, in order to see him the same way that A sees B' on the yonder side of the mirror?*

In every case, A must go perpendicularly opposite as indicated above (i.e., into the hinterland behind the mirror) so that he has the same direction of view toward *his* friend (B) as he had earlier toward the mirror twin (B')—of course, with the mirror now pulled out of the way (see Figure 12).

However, he also saw B' right-left reversed (compared to B). Think a minute about the situation that prevails at this spot—A'. This is precisely where B saw A's twin, right-left reversed in mirror space yonder. A cannot simply go behind

Optics

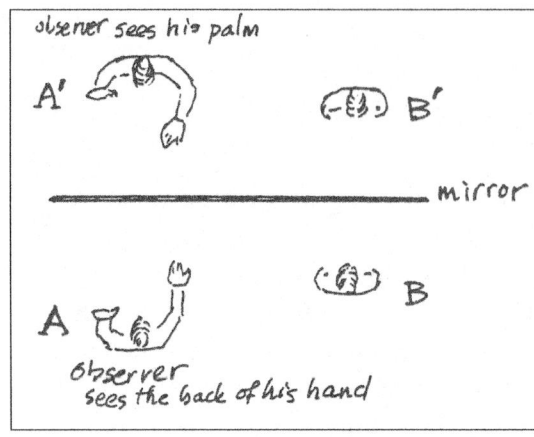

Figure 12. The corresponding viewpoint

the mirror, since that place in the hinterland (tangibly experienceable) is not totally equivalent to the yonder place (not experienceable): It also appears right-left reversed!

All the people visible in the mirror find themselves in a land in which people reach out to shake hands with the left and generally do everything with the left (turn screws, cut with scissors, write), as we always observe for these yonder people. If person A' yonder comes to greet B on this side with his usual hand (he calls it his "right"), then he sees a person on the other side who offers the wrong hand (we call it the "right") and B classifies A' as different than (his) norm, as a "left"-hander. But, how unhappy A' will be when he observes that all the people yonder (for us they are on this side) use the wrong hand, the wrong foot, etc.

Before we set aside the bewitched mirror (with a sigh of relief), we should express these experiences in a rule, the concluding mirror-law: *The view beheld by the yonder side observer of things on this side is equivalent to the view that an observer belonging to this side has of the corresponding things in yonder space.*[8] I see in the yonder side scene the same view of things belonging to this side as my mirror-twin sees of corresponding things on this side, from his viewpoint on the yonder side.

If we shorten this to express the connections in purely visible—albeit general terms—we get the Mirror-Law. It cannot be formulated exclusively for this side. It has value only for those who want to exercise their reason on what they themselves see. Usually we try to retain the same line of thinking in the Mirror-Law, whichever side we wish to take as the reference.

Konrad Witz (ca 1400–1445), the first landscape painter of the European tradition, has a similar mirror law be in his painting *St Christophorus* (in an uncertain restoration, in the Kustmuseum Basel). He painted the church reflected in the water with the perspective if he were looking from the position of the monk (painted in the scene), gazing into the water. This view looks diagonally upward to the church, as if up from under the pond, from the so-called yonder side.

The right-left issue is also treated by Michelangelo in his Sistine Chapel painting of the *Creation of Adam*. The relation of going forward to resistance or hesitation, of spirit to soul, of masculine to feminine, and of lung to heart is hinted at in this and other of his works—at questions about the goal and origin of mankind. What right-left laterality fundamentally means for the human being can only be vaguely indicated here.

One final indication: For someone with right-left reversed (i.e., a mirror arrangement of dominant sides), if their concealed laterality is taken as the whole reality, we do injury to them with respect to their future-creating personality, i.e., in their temporal development. Also, an educational treatment, readable by students and giving a therapeutic approach to reflections, is found in the children's story by Fynn (aka Sydney Hopkins), *Mr. God, This Is Anna*.

4. CRITIQUE OF THE YONDER SIDE: TRAINING OF PERCEPTION OF IMAGES

The term "yonder side," which I have introduced as a kind of mythical analogy, must again vanish from optics! It seemed helpful to me to aid the reader's grasp of the visual reality of the world, since today, people think more in concepts than in perceptions. To bring our thought into contact with the perceptions, one cannot be too radical. But, aside from its rhetorical introduction, the word "yonder side" is inadequate not only because of its ambiguity, but because the mirror generally does not open into anywhere (except perhaps to our sight). Much more, it forms a barrier as a surface interposed in our line-of-sight.[9] As we have already presented in Goethe's poem, the mirror does not bring a new view and direction to the images, but rather a freezing of the overall picture. It does not animate, but puts braces around things. It is itself an immobile and dead surface.

We can find an account in the Persian Mysteries of how there is a counter-world to this world, a right-left reversed world, with an autonomous existence—it was created separately. It is not oriented forward (right) but backward (left), and it invisibly permeates this visible world. That world is made visible in mirrors, i.e., in the water's surface in the Persian temple. This is where we first learn of the idea of the mirror as a gateway to a counter-world. Still, the physicist remains undeceived. Such a legend has to do with interesting facts, from a spiritual viewpoint; but, for the modern way of perceiving nature, it is valueless, as it deems our perceptions in a mirror to be the product of some mythical yonder-world. Instead of presenting the things connected with mirror images as something to be researched with theories of vision, it presents these phenomena as a product of something beyond science, created by the gods. We gain nothing in understanding if we just say "That's just how it is, the way it was created once upon a time."[10]

In actuality, the mirror does not open on to anywhere else. In reality, it presents the world as one-sided and distorted. This distortion occurs in three stages:

1. *Looking grazing the surface,* with the head held near the water's surface, a symmetrical scene appears. (See Figure 7, fingers at the edge of the mirror.) Visible things are drawn over into another image in the mirror.

2. *Looking obliquely* (at about 45° compared to the earlier view), the observer is drawn further into the mirror-space (termed the yonder side, above). Mirror image and object are more distinct and separated. Moreover, the objects visible in the mirror are reversed, have different portions visible, and are seen from a totally different direction of view. This drawing-apart and reversal of direct view is typical for the mirror.

3. If I *look face on* into the mirror, perhaps at my outstretched hand, the view is not just shifted now, but independent of what is visible on this side. There, I see views which can never be seen on this side, e.g., my own face; instead of the backside of the hand, I see the palm. And, what is usually seen in the mirror, can only be seen on this side if I don't look into the mirror, but turn all the way around and look at the scene behind me.

The shifting and separating of images into two can be seen most clearly in the second case, where the observer himself is in the scene. Also, there we don't have an entirely different scene, as in the third case, but just an increased shifting of the image. The third case is connected more with the well in the fairy tale of Mother Holle—with an entrance shown by a mirror into the realm of the clouds.

Thus, we have three mirror scenes that we can observe carefully. In organizing these variations of what we see, one sentence in Rudolf Steiner's curriculum plan is helpful: "When you seek connections in this way, no longer explain the mirror phenomena by saying: That is a mirror, and a ray of light falls vertically upon it, and we have here the eye and will have to explain why, when the eye sees directly, nothing further happens other than that we see directly. You must come to see how, taken fundamentally for the eye, the mirror shifts (pulls over) the image of the object."[11] The main thing is a transformation of view (image), which we wish to penetrate with thought, rather than using light rays imagined (in a mechanical metaphor) as striking the mirror, and bounding off according to some rule. Mirror images should not be understood—as is our custom—as being physical occurrences in space, but rather as an experience worth studying for its own sake, the study of which strengthens our connection to the world.

III. PHYSICS IN SCHOOL

1. TRANSITION TO CONVENTIONAL MIRROR-LAWS

The study of a transformation of images, in contrast to reflection of a picture unchanged, is a challenging exercise for the intellect, especially for an atomistic, intellectual used to thinking in terms of light rays. Pedagogically, our path is strikingly different.

Here is a thoughtfully worded formulation of our "purely visual mirror-law" for a naïve understanding on first meeting a mirror: Things seen in the mirror (on the yonder side) can be traced back to what is present in front of the mirror (this side); we imagine each visible point is repeated perpendicularly across the mirror, i.e., all objects, present in front of a mirror are also present (mirrored) in the yonder space, they appear over there in a normal way from our viewpoint here. Buildings that lie at the mirror plane become symmetrical, like baroque castles with two identical halves. We can call this the "perpendicularly opposite and equidistant" law.

As I move along the mirror, the line-of-sight to the half of the building seen beyond the mirror changes. Niches in the facade become invisible, just as they would for a house really standing there. If objections are raised, we can add to our experiential, "purely visual mirror rule," the more geometric mirror-law above. The price of this simplification is that material concepts are put in place of interconnected visual perceptions. In an imaginary way, we project tangible objects into virtual space, behind the mirror.

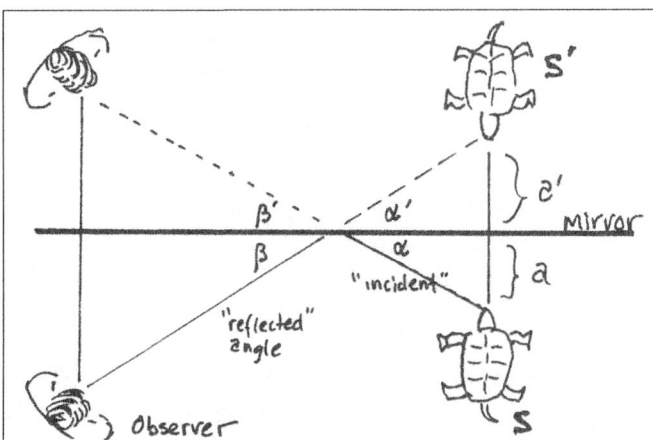

Figure 13. Perpendicularly opposite formulation of the mirror law

However, from our perpendicular-opposite law it is an easy step to the conventional law: Angle of incidence = Angle of reflection.[12] From a bird's eye view of the perpendicular-opposite law (Figure 13), the lines A-A' and S-S' are perpendicular to the mirror surface, a = a' and also ∠a =∠a'. Since b is an opposite angle (vertex angle), we know ∠b =∠a; the angle of incidence a viewed from S is the same as the exit angle b, viewed from the observer A. Since he knows nothing of the deflection of the line-of-sight, he takes the turtle to be at S—behind the mirror. This holds for any other animal on this side, and so holds for all points of scenery on this side, creating a yonder counterpart for all points. In expressing the law of angles of reflection, rather than referring to "rays of light" it is better to speak of "line-of-sight" (direction of view, where things are visible or obscured).

The equal angles of reflection law can be confirmed by the fact that two people can see each other in the mirror. If the exit angle of the first one was greater than the incident angle of the other, then this would also hold for their view looking back the other way; they would each be looking in a different direction (steeper) toward a different spot on the mirror. The possibility of mutual seeing in reality proves the opposite: incident angle = exit angle.

2. EDUCATIONAL DIDACTICS

The lessons about the mirror (over 3 days in grade 7) may be introduced with the pool-reflection mirror and mirror manufacture experiments (Op1a & 1b). Following that is experiment with candles and shadows (Op2a) where we investigate how the shadows fall across the border. As a third experiment, we do the alternate observer positions experiment (Op3a). If this rule is too hard for them to work out, they can usually grasp the simplified "perpendicular opposite" rule (see above).

Here is a nice supplemental experiment: Challenge a student to write his name on a piece of paper, so that when seen in a vertical mirror (while the paper itself is concealed), the name will appear correctly in the mirror. Or, while looking in a mirror, draw a rectangle with a diagonal slash on a piece of paper on the table. Such activities help the students intensively experience the mirror's distortion.

Another tricky exercise is to solve the problem: How large must a vertical wardrobe mirror be so that a person can see himself from head to toe? What effect does the distance from the mirror have? We will see ourselves perpendicularly opposite, as if through a window into mirror-space. The answer is independent of the distance of the observer—the mirror must be half as tall as the observer [because the mirror will always be halfway between the observer and his mirror "twin." This is the law of similar triangles.].

Another problem question: In front of a mirror wall runs a curtain. Where must a crack be drawn open so that someone—from where he stands—can see an object standing in some other arbitrary spot? Or, a more difficult question: Under what conditions could someone see in a mirror an object present in their vicinity, which shows its yonder location, but whose position on this side, if I turn about, cannot be determined? All of these exercises can be solved without recourse to light rays, by using our perpendicular-opposite law. Other applications such as the dentist's mirror, the kaleidoscope, and the periscope (an angled tube with a mirror in the angle) are well-known.

The concept of the so-called "yonder world" may be used only provisionally in the grade seven block, and mainly as a stimulus to marvel over the mirror initially. A thorough evaluation of it, but with the full participation of the student's own world-view, will be necessary in the grade 12 physics of light block. There, confronting the most general forms of the mirror law, the students are presented with a most stimulating question: Can I master these problems with my dull arsenal of thinking tools, or am I too weak in spirit? Also in grade 12, the experiential-method of thinking will be contrasted to the ray-model.

In the preceding discussion, more material has been presented than is absolutely needed for grade seven. The objective has been to show that a Goethean method, which penetrates deeply into the senses, can lead to the same mastery of the phenomena as the ray model Moreover, it fosters a concrete and strengthened thinking and thereby fosters real development. Goethean optics does not mean generating childishly pretty pictures. Rather, it means thinking energetically and in a way that is fully present.[13]

IV. STEPS TO UNDERSTANDING THE CAMERA IMAGE

After our study of the mirror, we turn our attention to the pinhole camera. I don't like to present the students with a finished pinhole camera experiment, but instead prefer trying to develop stepwise the elements of the camera through various demonstrations.

1. STYLIZED SHADOWS AND COUNTER MOVEMENT

Indeed, the shadows in a sun-filled landscape show it all. That is to say, as we have already observed with shadows, the sun remodels every shadow scene in its own image (see, Grade 6, section III.3.c). The further the shadow is from the object, the more the shadows become roundish, indistinct forms. The tiny openings in the canopy of leaves in a forest never reproduce their polygonal shape in the light spots on the forest floor, but show up as circular discs. The irregular apertures are recast toward circles, i.e., they approach the shape of the sun. Shadows, as well as bright spots, take on the form of the prevailing self-bright light. This can be shown in the demonstration using a lamp and crescent-shaped mask (Op4a), in which we cast a crescent-shaped light spot using a shaped shadow-mask. (Compare grade 6, section III.3 and Op10.) A second important effect is that the crescent-form occurs inverted: A bowl becomes a cap and the inverse. We can verify this for the children: Move a roundish leaf around in a light-beam, while looking at a canvas (a very good projection surface) on the other side of a bush or some latticework. The lamp becomes a partially visible crescent, and the dim light spot on the projection surface takes on the form of an inverted crescent.

This constant inversion of the lamp shape, the formation of an image on the opposite side that is reversed, can be demonstrated in another large-scale experiment. Behind a set-up of shading panels (Op4b) moves an illuminated white form (e.g., from west to east). On the darkened side, a whitish, vertical stripe moves along the projection wall in the opposite direction (e.g., east to west). Out of such demonstrations of reversal, we can clarify the inversion aspect of the crescent experiment above. Also, this counter-movement is strongly based on a principle that we already utilized for shadows (see grade 6, section III.3): We get brightness at those places from which we can see the bright source, and we get darkness (shadows) where we can see only darker portions of the surrounding cupola. The bright strip is always located just at the place from which we could see the brightly illuminated figure looking back through the crack in the panels. We may recognize this principle also in the light and dark stripes running across the far wall when there is a crack in the curtains, and people or autos move past outside.

2. THE PINHOLE CAMERA (CAMERA OBSCURA)

Quite often in an almost dark attic or in a cave, we can see a ghostly image cast on a smooth wall: The landscape outside in front of the cave appears inside in indistinct blotches on the wall opposite the opening. People move about on their heads, with the sky beneath them. In addition to seeing the pinhole scene inside the cave, we see that right & left are inverted—although this is less noticeable than the overall upside-down orientation. We can produce these phenomena with a darkening curtain (Op5a). The explanation is in front of our eyes. If an observer walks slowly along the projection surface (the far wall), he will see only tiny bits of the landscape outside—if the aperture is very tiny. He can hardly know what part of the outside scene he sees, but it does possess a definite brightness and color, different for every point on the projection wall. The projection wall acts as a roving eye to sense the light & dark, manifesting it on its surface (metaphorically, of course). With such thoughts, we can relate the pinhole camera back to the "all-seeing globe" in grade 6 (Op3). That such an image of the whole occurs, right down to details, remains a marvel.

Self-evidently, the pinhole image is dim; we should become clear about the phenomena in this regard. As we make the opening larger, the image gets brighter, but correspondingly less sharp. The same thing happens if we decrease the distance between the opening and the projection wall. The pinhole image occurs just before complete darkness, when it is completely sharp. The same holds true for shadows: When the shadow-casting object lies directly over the shadowed surface, the shadow will be completely sharp. The accuracy of the shadow, i.e., the way the shadow is cast, occurs mechanically on each side of the illuminated region when the surface falls out of connection with the brightness that weaves throughout free space. The imaging of the light is much more a transformation of images than a mechanical copying; it brings new images into existence, interwoven with the whole surroundings. This also occurs with sun shadows, and also with the counter-movement of multiple shadows. Such multiple shadows occur when two objects cast shadows and one is further from the shadowed surface than the other and then one is moved transverse to the shadow alignment. That occurs in nature with clouds, trees, bushes, etc. Here we have the archetype of the pinhole camera. Shadows do not fall mechanically on one another, but transform their shapes depending on their proximity to the light, etc.

Homemade cameras are very popular with the students. We give some guidance as to how they can be made out of a tin can (Op5b) or cardboard (Op5c) with parchment paper or tracing paper on the end. If the teacher wants to introduce such a camera in the classroom, then he can throw a dark cloth over the student as he or she looks into it and let just the front part poke out from under the cloth. Then have the observer report which students in the classroom are standing up, where they are, etc.

Experiments in Optics

OP 1 POND-REFLECTION MIRROR

Uncover a large (75cm / 30" square or larger) dresser mirror laid horizontally on a table in front of the class. Hold a magazine picture of a landscape behind the mirror so the class can see it and also compare its mirror image. [This is our type one or watery mirror situation; see above discussion for the nature-connections. This always shows a symmetrical pair of images; the colors and shapes are the same, but the mirror counterpart is flipped over or inverted.]

Have a student move his fingers up over the edge of the mirror. Note how the mirror fingers appear essentially the same but inverted, and move downward. [Drawing this in the main lesson book makes a good exercise.]

OP 1B GENERATION OF A MIRROR-SURFACE

Sprinkle the horizontal dresser-mirror with white dust (talcum powder, flour or chalk dust) and set up a lamp above and slightly to one side to illuminate it. Darken the room and turn on the lamp. The mirror appears as a white surface. Now, begin to clear away the powder, making the surface smooth (analogous to laboriously polishing the roughness away with a rouge; note how the improved mirror becomes darker and even invisible, while the surroundings (the ceiling) become more visible.

Have the students come up and look into the "hole" we have cleared away in the powder. Now the mirror appears as a transparent "window" into a mirror room "yonder." Position a hand over the mirror window and note how a mirror counterpart hand appears. Note also how both hands behave normally and cast shadows—the this-side hand casting one onto the mirror ceiling and the yonder-side hand casting one onto our ceiling! Also, note how our point of view is different toward the yonder space, e.g., we see the palm of the yonder hand but see only the back of the hand on this side.

OP 1C DIRECTION OF VIEW

As we move to various viewing positions, notice how the region we see (as if through a mirror window) also shifts [further reinforcing the quality of the yonder space as a portion of a visually "real" yonder-world].

Experiments in Optics

OP 2A DOUBLESHADOW SCENE

Position a smaller rectangular mirror vertically in front of the class, on a white paper- or tablecloth-covered table [the mirror-on-the-wall or window situation]. Place a tall candle about 20 cm (8") away from the mirror and to one side. Place a smaller figurine closer to the mirror and to the other side. Note how there is now a mirror counterpart candle and figurine "yonder." Dim the room, and light the candle. Observe how it casts shadows that fall across the border between the space this side and yonder-side without disruption. Note how the yonder-side candle also casts a very effective shadow, which falls across the border unbroken and runs out into space on this side! [Visually, the yonder scene—as well as the well-known things here—are three-dimensional regions; both act properly in casting shadows, with closer objects obscuring more distant ones (parallax) as we shift side-to-side, etc.].

OP 2B MIRROR "WINDOW"

Explore how the scene visible in the vertical mirror-"window" is not quite the same as the scene on this side. If possible, set up a large unframed mirror outside and sketch the landscape, including the mirror rectangle and the yonder landscape seen within the edges of the mirror. Look at various objects: conch shell, screw, watch face. Note how they have all the parts, but they are inverted or switched. Sketching the pairs of objects is challenging but is a good exercise in observation.

OP 3 PURELY VISUAL LAW OF MIRROR SPACE (THE BOX YONDER)

Place a student and a figurine in front of the vertical mirror and ask him to locate the place where he will have the same view of the figurine this side as he now has of the one yonder side. [He must stand at the position where the rest of the class now sees his mirror twin beyond the mirror.]

OP 3B TRANSITION TO CONVENTIONAL LAW OF REFLECTION

Show various shapes or names written so that they appear correctly when viewed laid flat in front of a mirror; compare to the original. What is the rule for drawing these? [Duplicate each point perpendicularly opposite the counterpart and equidistant from the mirror. This is a more abstract, geometric restatement of the experiential rule derived above.)

OP 4A STYLIZED SHADOWS

We use a cardboard box with a round window (grade 6, Op6 & Op9). A round disc the same diameter as the window is fastened to the front, so that 1/3 of the hole shows as a crescent shape. The carton is placed before the front row of students near the side wall, with the opening facing the blackboard. In between it and the board, we put a bush or branch with moderately small leaves (e.g., rhododendron). As we darken the room, the bright bulb inside the carton casts a shadow of the leafy branch on the board. We show the students the crescent shape of the opening (crescent, points upward like a balsamic moon), then turning the carton back. we observe the light splotches between the shadows of the leaves. Then invert the box so its crescent window stands pointing downward. Finally the students notice that, between inversions, genuine though indistinct crescent shapes are visible; they invert when the box is inverted. However, they are always oriented 180° from the way the crescent on our carton lamp is pointing.

Young people may need to see this several times to get it clear. The sharpness of the crescent shapes also depends on the type of foliage and on the distance from the light source; the teacher should experiment to get the best combination. An experiment is most suggestive, that is to say is a stronger esthetic experience of the whole if it remains somewhat inexplicable, at least until we see what follows.

OP 4B COUNTER-MOVEMENT

We build the following stage flats in the front of the classroom (see Figure 14). The baffle that shades the lamp from the class could be just cheap cardboard; the screens could also be made of large blankets, hung over map stands or on a clothesline. Have the student subject walk to & fro; the light strip shifts opposite: fro & to! [Student subjects with brighter clothing will produce the best results, naturally.]

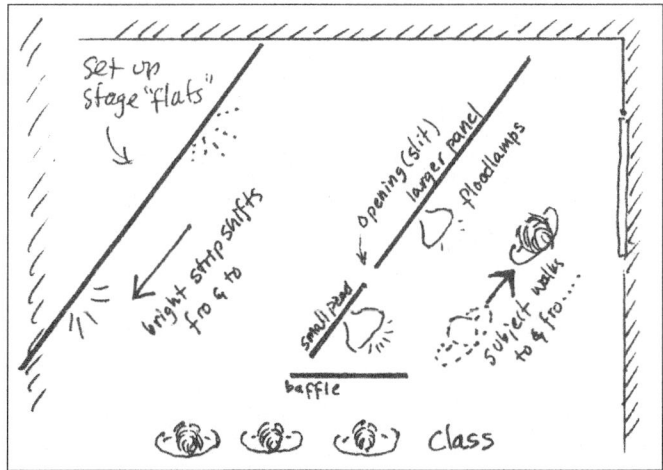

Figure 14. Counter-movement demonstrations

OP 5A CLASSROOM CAMERA

The more completely the room can be darkened, the larger a pinhole picture we can produce. The projection screen should be at least 5m (15') wide and visible to the whole class. An adequate substitute is a slide projection screen set up on the lecture table in front. The hole in the curtain (or shade) should be no larger than a fingernail and positioned approximately at eye height, depending on the landscape outside. If it is lower than the windows, then position the hole a bit lower. Before we start the demonstration, we get one student to go outside and ask him to go back and forth in front of the window, running one way, strolling the other, hopping on one leg, holding one arm up, etc. Also, we ask him to report what color the hole has from outside, whether it changes in brightness during the experiment.

OP 5B TIN CAN CAMERA

The students really like to build their own pinhole cameras. The simplest can be made out of a tin can. We cut off the top and fasten parchment paper (or tracing paper) to the top end with tape. We puncture the bottom with a can opener, leaving a hole no more than 1–2mm (1/8") in diameter. The image will generally appear very weak in a daylit room; window framing or ceiling lamps show up best.

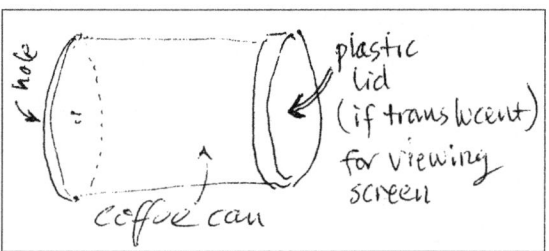

Figure 15a. Tin can camera

OP 5C CARDBOARD CAMERA

More effective is a camera built out of cardboard with a cutout for the face. The outer box with the pinhole (left, in Figure 15) has a sliding inner box with a translucent window (use tracing paper or oiled typewriter paper) and a cut-out for the nose and forehead:

The eyes are kept in the dark so more details of the image are visible. Also, we could measure the degree of reduction of the image by removing the front and sketching lightly on the paper with pencil the outline of the scene we saw.

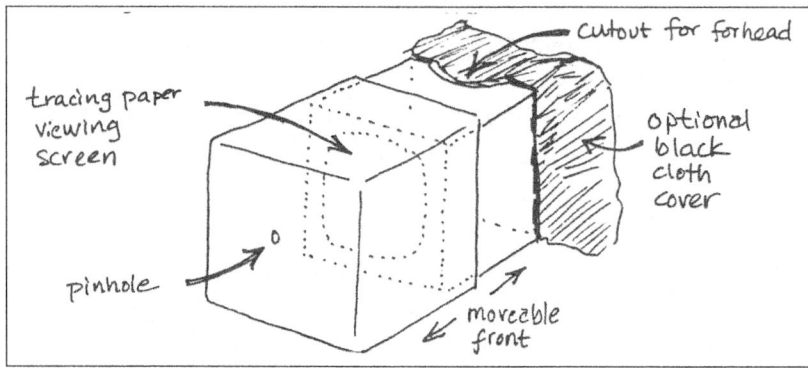

Figure 15b. Cardboard camera

Grade 7 Heat

INTRODUCTION

In the seventh grade, the lessons on heat seem much easier to present since they begin to utilize and implement conventional concepts. However, these concepts must be developed with clarity. The topics covered include expansion, the temperature scale, and conduction of heat.

I. WARMTH AND SURROUNDINGS

Using a pan of hot water, bundled up in various materials, we demonstrate the amount of cooling which occurs for various commonly available materials (He1). For example, when packed in wool, the water remains hot although the wool itself doesn't get particularly hot. It produces an island of warmth, separating the water from the surrounding conditions in the room. Such materials which thermally separate (isolate) are termed insulators. Still, a cold or warm object can never be totally isolated (insulated); even the best thermos allows ice to melt, or coffee to get tepid after a few days. We inadvertently demonstrate insulation in grade 7 chemistry, when heating a chunk of diatomaceous limestone. Although red hot on top, the bottom can still be held comfortably in our hand. One group of materials which doesn't insulate, but rather allows warmth or cold to equalize out rapidly, is the metals. They are nearly indistinguishable in the speed with which each one allows the other end to get hot when one end is heated (He2).

To see this in practice, imagine a large tank with water kept at boiling temperature. Another with cold water adjoins it through a common wall, made out of various materials. The cold water will get hot at a distinct rate—depending on the material. The cold water tank is otherwise very well insulated, so it is heated only by the common wall, and not affected by the surroundings. We imagine the common wall initially made out of felt, 3 cm (1-3/16") thick, coated with rubber to keep it dry. A moderate amount of time elapses until the cold water has been heated and its temperature rises measurably.

Now, we remake the wall out of a layer of cork. In order for the cold water to remain cold an equally long time, the wall must now be about 3.3 cm thick. For other materials, it must be even thicker (see table below).

Heat

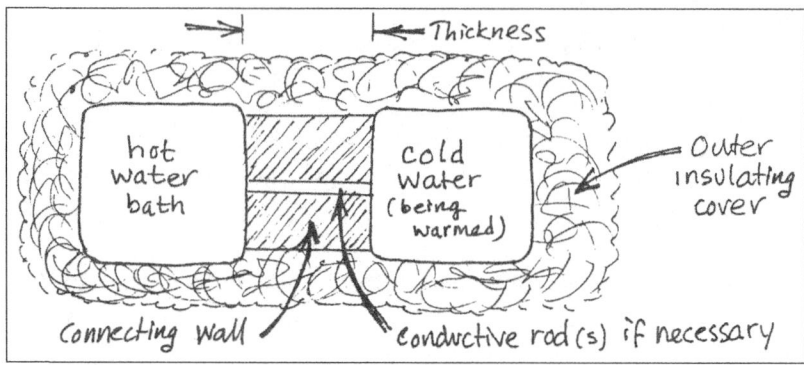

Figure 16. Insulation set-up

These insulation values are easily determined only for the poorer conductors (through granite). When the imagined wall between the tanks would have to be meters thick, its lateral surfaces should be insulated. In addition, the hot end of the connecting material would have to be heated, or else the water would initially get warm too slowly.

Relative Thickness Producing the Same Warming

MATERIAL	THICKNESS
Sulfur	1 cm?
Cork, horsehair, silk, wool felt	3
Wood, sawdust, diatomaceous earth	13
Brick, porcelain	40
Glass	130
Stone: granite, sand	290
Iron (steel)	6100 cm or 61 meters
Aluminum	204 m
Copper	342
Silver	428

The astonishing length of nearly 1/2 kilometer of silver needed in contrast to 1 cm of sulfur would be difficult to measure directly in an experiment. For such materials, it would be better to make the connection using only a few rods of metal, embedded in an insulating material. The contact surface area is thereby reduced to the sum of the cross-sections of the rods. For the same material, the warming is proportional to the contact area. Thus, after a few calculations, we would find that silver would permit heat to behave as if the tanks (actually separated by a great distance) were in close contact. For warmth, silver equalizes differences quickly, sulfur separates. We see which materials isolate something warm or cold from its environment, as if creating a great distance between it and the surroundings, which is not physically there at all, and which materials unite things with their environment.

With the concepts of *insulator* and *conductor*, we can now describe the heating system in a building. It will work best if we construct a drawing from a tour of the components: furnace, boiler or heat exchanger, hot water (steam) pipes or hot air ducts, radiators. In the boiler, the water and the flames are separated by only a thin (4 mm) steel wall of the firebox. The fire chamber is insulated from the basement room by fire-clay tiles and brickwork. The pipes in the basement are padded with glass wool. In the living spaces, in contrast, there is only a very thin wall (1 mm) of sheet metal between the water and the air, and the radiator is constructed so its surface area is very large. Students can describe each of these insulating and conducting components along with their drawings.

It is never possible to prevent a warm object from eventually warming the surroundings or making them participate in its different heat state. Without the sun, without living beings, volcanoes and fire, all differences in warmth would equalize out and vanish. *These variations in warmth are the expression of the life of beings and of the earth in its cosmic rhythms.*

II. THERMAL EXPANSION

We begin studying expansion with SOLIDS. At first we show the metal ball, which, after heating, no longer passes through the ring (He3). A bit more impressive is the shattering glass disc (He4) or shattering bolt (He5). The fact that all solid materials expand differently is very characteristic of this kingdom of rigid, dead materials. Incidentally, these variations are shown in the expansion race experiment (He6), and the following table may be enough for now:

MATERIAL	LINEAR EXPANSION*
Aluminum	0.25 mm
Copper	0.17
Iron (structural steel)	0.12
Concrete	0.12
Glass	0.09
Quartz	0.005

*Increase in length of a 1 meter rod/10° Celsius

Heat

FLUIDS also have distinct and generally larger expansion coefficients, and are arranged according to their increase in volume. We demonstrate this cubic expansion using a 2-liter flask with a glass stem to make the expansion visible—which is our travel thermometer (He8). It is astonishing how quickly we see an effect for a barely perceptible warming. From the marks on the expansion tube, we can directly read off temperatures, if it is calibrated in reference to standard temperatures (such as boiling water, body heat, ice water, etc.).

MATERIAL	VOL. EXPANSION*
Water	2 cm^3
Mercury	2
Alcohol	11
Toluol	12
Iron (see above)	0.36
Air	30

*Increase in volume of 1 liter/10° Celsius

GASES all show the same very large expansion coefficient (1/273rd of their volume) per 1°C, so the investigation of one gas—the air—may suffice. We show the volume increase on the one hand, and also the increase in pressure (He9) on the other. Gases are compressible (e.g., a bicycle inner tube). In filling a container with liquid, a small air space should be left because liquids are nearly incompressible, and otherwise they would pop the cap off when heated. Expansion of solids can be an image for us of the spreading out of heat. Characteristically, gases are most strongly affected by this effect.

III. THERMOMETERS

To fill a thermometer, we require a liquid (since it will not compress), something that will show the greatest expansion for the same increase in heat and which also will not boil quickly or freeze. In addition, it should be able to take up the warmth of its surrounding quickly, i.e., it should be a good conductor. These requirements are best met by mercury. Grain alcohol (ethanol) also has a large expansion coefficient, but a lower conductivity. Alcohol, pentane or toluene are particularly useful for low-temperature thermometers, since mercury freezes at -38.9°C (it boils only at +357°C).

To construct a liquid expansion thermometer requires a scale, a separate, totally enclosed liquid space including a bulb (which is affected by the heat) and a capillary (a narrow diameter tubing, which makes a given expansion in volume show up as a large vertical rise up the tube). The evacuated liquid space is filled with nitrogen at one atmosphere, to prevent an oxide film from later forming on the mercury.

Another enclosed type of thermometer has the scale behind the thin capillary tube, all enclosed in an outer protective cylinder. A simpler type has the scale engraved directly on the thick-walled capillary tubing itself. But, with the greater distance between the scale and the mercury capillary, this type is harder to read accurately. A nitrogen or "high-range" thermometer is filled with nitrogen at 50 atmospheres pressure, so the mercury will boils at only 675°C. Such a thermometer is usable up to 600°C.

The sensitivity of a thermometer depends on the capacity for expansion or "expansion coefficient" of its liquid fill. The construction also affects the sensitivity: the capillary should be as thin as possible and the bulb as large as possible (as great a liquid capacity as possible). Thus, a large increase in volume and a great rise in the height of the capillary column will occur for a given change in temperature. For this reason, a great sensitivity is also exhibited by our water thermometer (He8). These kinds of mercury thermometers do not have an accuracy exceeding 0.01°; for usual fine measurement, they are made with scale divisions of 0.1°.

Calibration requires at least two reference points in temperature. In simplistic terms: We use the melting point of ice, and mark this as 0 degrees. Then boiling water at normal atmospheric pressure and at sea level is marked as 100° on the Celsius scale. This scale was developed by the Swedish geographer and astronomy professor Anders Celsius in Uppsala around 1742.

The instrument maker and researcher Gabriel Fahrenheit of Amsterdam was the first (1715) to make thermometers which were calibrated, i.e., indicated temperatures which agreed with each other. He utilized three fixed points. A chemical cold-mixture of brine (water & table salt) and sal ammoniac (ammonium nitrate salt) registered the lowest position on the capillary and was marked as zero. He chose the body temperature of a healthy person as another referant, and marked it as 96°.[1] (Since this is divisible by all small numbers except 5 and 7, it is easily subdivided.) On his scale, the melting point of ice falls at 32, and boiling water falls at +212° F.[2]

IV. NEW SORTS OF ICE EXPERIMENTS

Under this topic of heat, it is interesting to study ice again, i.e., in cold-mixtures (He10). Along with hardening, salts increase the capacity of natural things to be separate. This tendency toward separateness is opposed in a solution by water. With the artificial melting (promoted by water) that occurs thus in the absence of external heat, this hardening-salt influence now manifests itself by the solution becoming cold (the hardening effect of salt is expressed in another way).

This can be connected with other phenomena involving cold water, e.g., the 4°C (39.2°F) density anomaly and stratification of freshwater lakes, etc.

One simple experiment, among many possibilities, is done with a 250 ml Florence flask, a thermometer, and a capillary tube (2–3 mm dia.) inserted through a stopper at the top, or with any large polished glass container, suspended in a brine bath (ice and salt water) which is kept stirred. Also, we could perhaps discuss phenomena from nature such as the freezing of salt water to ice (see polar explorers' biographies). At about -2°, seawater finally begins to form a granular ice mash. In order to get it this cold, the weather must be intensely cold, since water's greatest density does not occur at once, but already at +4°C. So, as water is cooled beyond this point, the colder & denser water does not stay at the surface to be cooled further, but sinks slowly downward, away from the frigid winds. Sea ice is turbid and riddled through and through with tiny bubbles and crevices, which contain the residual enriched salt water, separated out from the portions which do freeze by -2°C. Thus, the phenomena of ice formation are manifold.

Experiments in Heat

HE 1 THERMAL CHAMBER

We use four 1-liter (1-quart) tin cans and four large glass pans, in which the cans are set, resting on porcelain dishes to insulate against thermal shock. Plastic pans could also be used, as they too are poor thermal conductors. Each can is filled 3/4 full with boiling water. In the first pan, the can is surrounded with air. Immediately after pouring the hot water in, we have a student investigate (perhaps with a thermometer, eventually) the air in the basin, the walls of the basin and briefly test the water in the can. The second pan is filled with cold water, and a cardboard cover is placed on top. The third pan is filled with fine sand, and the fourth with felt or wool, also in a layer under the tin can. After about 5–8 minutes, have students use their fingers (gingerly) at first and then use a thermometer to examine (in the same sequence) the contents of each tin can, the pan at various distances from the tin can, and the exterior wall of the pan. Establish which material forms the greatest thermal barrier (insulator) and which the least. Discuss practical applications such as the heating system, as outlined above.

HE 2 HEATING RODS

We use various metal rods, at least 30 cm long and 1.5 cm diameter: copper, aluminum, steel, and brass, and glass tubing.[3] Show the class that each rod initially has the same degree of coolness over the whole length of the rod. Set up five small Bunsen burners on the demonstration table, using Y-connectors and rubber hoses to the gas supply. Adjust the burners to small blue flames and show the students where to hold the rods at the tip of the inner blue cone.

Heat all five rods in the flames (use care with the glass tubing to prevent breaking). After a minute we can already have another student examine the rods and see which parts of each rod have already become hot and which are still as cool as before, or else use a wet finger to see where it will sizzle on the hot parts. (Show how a wet finger sizzles everywhere on the copper rod.)

The student holding the copper or brass rod will be the first to want to drop his rod when it gets too hot to hold—instead set it on a lab tripod so it can be kept in the flame during the whole experiment. Next the aluminum, and finally the steel rod will become too hot to hold and are also set on tripods. The glass rod most often can still be held!

Experiments in Heat

HE 3 EXPANDING BALL & RING

The well-known ball & ring apparatus, available from science supply houses, is introduced. We first show how the untreated ball will pass through the ring. Then push the ball through the ring and heat it to a dull red glow (don't melt the shaft!). After heating, the ball will not come back out until the ring is also heated.

HE 4 GLASS DISC

An impressive (and possibly dangerous) experiment is to fracture a glass disc about 10 cm (4") in diameter using heat. But with precautions, the danger is minimal. A glass disc, not thicker than 3 mm (1/8"), is set up on wooden blocks so that the tip of a candle flame just touches it in the center.

After a few seconds, the disc shatters with a bang, and fragments spray out in the plane of the disc in all directions. Therefore, we put the candle on the floor, surround the experiment with a cardboard fence 2x the height of the disc, and have the students stand on their chairs. So the teacher knows what to expect and is well-prepared, experiment with this set-up prior to the lesson. A more likely danger over which we have problematic control is a student will attempt to repeat this experiment at home.

HE 5 CANTANKEROUS BOLT

Show an ordinary, large bolt & nut (say, ½" diameter) at room temperature, and see how the nut screws on and off easily by hand. However, when the nut is screwed on and the threaded shaft is heated, the nut cannot be screwed off even with pliers—until the nut is also heated [a practical application of He3, well-known to auto mechanics].

HE 6 EXPANSION RACE

Use two rods are used, mounted fixed at one end as in Figure 17, with the other end free to shift and resting on a needle with a paper straw pointer (the needle punctures the middle of the straw). When the rod moves, the needle rolls and the straw pointer turns, indicating the relative expansion of the rod. Set the straw pointer vertically to begin. The solidly fixed candles are lit on command and pushed simultaneously under the rods to start the experiment. In 4 minutes, using 7 candles, the aluminum rod will have rotated its straw 180°, and the iron only half that distance.

HE 7 EXPANDING PENNY

Give each set of students a votive candle, 2 old-fashioned double-edge razor blades (dulled with sandpaper) and 2 large clothespin clamps (or binder clips from an office supply store). Clamp the razor blades as indicated in Figure 17 so that the penny just passes through. After warming, it will not pass through, but when cooled, it will suddenly pass through again. The expansion is 3/100 mm per 100° (or about 1/10 the thickness of an extra-thin Pentel mechanical pencil-lead).

HE 8 TRAVEL THERMOMETER

A round (Florence) flask of at least 2-liter capacity and made of Pyrex is prepared with a rubber stopper and a 8 mm glass expansion tube. Fill it to the brim with ink-colored water and then insert the stopper with tubing. Warm it briefly, first just with hands, and finally

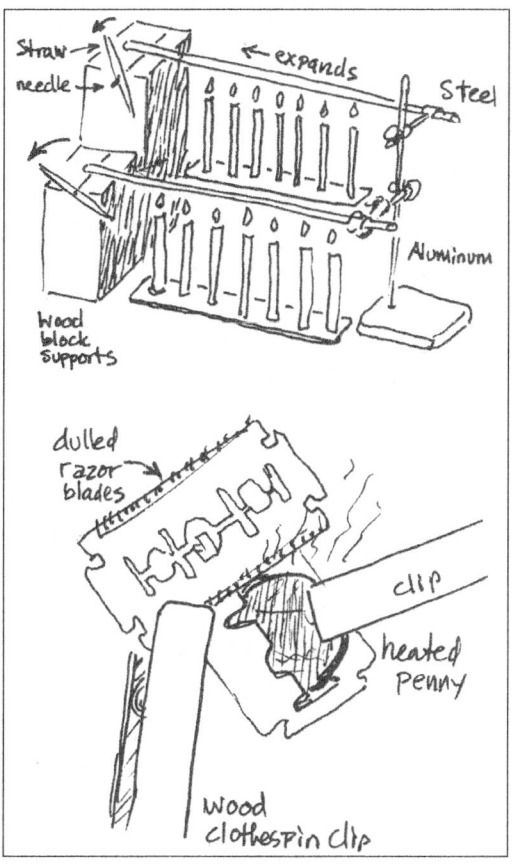

Figure 17. Expansion demonstrations

with a burner turned soft and low; observe the water climb in the tubing. Discuss how we could calibrate this thermoscope so it would read off the standard weather temperature. Galileo invented such a device and had the skilled glassblowers of Florence make it for him. These Florentine curiosities were the rage in every carriage for a while, to entertain travelers during their journey.

HE 9 EXPANSION OF AIR

a) Using the above flask, we replace the straight tube with a u-tube with a pool of colored water at the bottom. (See Figure 18a)

When the flask is simply held with warm hands, the colored water in the u-tube is pushed outward. Compare different students' hands. Discuss why this type of thermoscope would not be practical for use in weather reporting [only because it expands too much].

b) Now, add a stopcock (rubber tubing with clamp) to the end and heat the air bottle over a small candle flame. After the flask is hot enough to sizzle, open the stopcock: the puff of air escaping will blow the candle out! (See the study of heat engines in grade 9 for more technical applications.)

HE 10 FREEZING MIXTURES

Inside a towel, pulverize very cold ice by hammering it to a pulp. Use a plastic bowl in a picnic cooler to keep the powdered ice very cold; mix up the brine, using one plastic scoop of ice powder for 1/4 scoop table salt; this brine mixture melts gradually and cools to -20°C (-4°F).

This can be nicely demonstrated to the class if we give each student a test tube full in a beaker containing a little cold water (Figure 18b). After dipping the brine-filled test tube in cold water and holding it quietly in the air for a minute, the tube forms a rind of ice on its exterior. Pour out the brine and allow the ice-ring to warm slightly so it can be slid off gently.

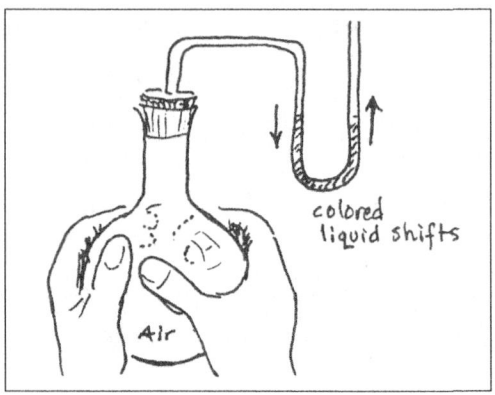

Figure 18a. Gas expansion

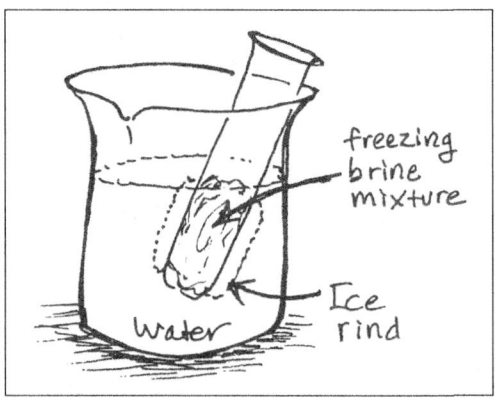

Figure 18b. Cold mixture demonstration

Grade 7 Electricity

INTRODUCTION

By the 17th and 18th centuries, the new vistas of research into electricity opened up by William Gilbert constantly produced new levels of discoveries, such as the electrification (charging) machine, the Leyden jar, the electric clock by Alexander Bain, the electric telegraph first by Georges-Louis Le Sage and later Samuel Morse, flickering glows within evacuated glass tubes, and finally the lightning rod of Benjamin Franklin. But, from this arena of static or frictional electricity, there never arose real electric engines or applications with practical value. Everything was toys—up to the lightning rod (which was only first utilized in 1754)—playthings which did not harness electricity for work in any sort of apparatus, but kept it at a distance. In this age, electricity didn't power any trades or tools; it didn't appear in any practical form in the home or in the workplace.

In the Baroque era, no one studied electricity. For the public at large, it remained a curiosity. Cavaliers and ladies of noble houses used it in bawdy social gatherings for entertainment. In an electrified cabinet, they charged up toy people in order to enjoy their nervous, spooky antics. In 1730, Stephen Gray first suspended a boy on silk and then touched his foot with a rubbed globe; paper scraps placed near his head suddenly flew toward his face. Alternatively, a chambermaid would be attached to the electrification machine while standing on an insulated stool; her "electric kiss" was a favorite initiation at social gatherings, but a slightly painful game for both parties. The first lightning rod was put into service in Paris in 1752. Still, Franklin's lightning rod didn't bring about any direct uses of electricity—he worked for electric protection not new applications.

The impetus for a new orientation which included practical discoveries was provided by the years of research by Luigi Galvani, while the epoch was actually brought in by the excellent research of Allesandro Volta. With Volta's work, a new, very primitive but measurably productive source of electricity appeared: Volta's "pile," or "galvanic plates," mounted in salt solutions (wet cells, the predecessor of the modern battery).

Electricity

I. GALVANIC ELECTRICITY

All sorts of old wives' tales circulated during the course of these discoveries. Many investigators had experimented in similar directions for a long time, especially with electric fishes. The so-called "crampfish" was well known, though no one thought of any electric explanation for it. The torpedo fish (*Torpedo marmorata*), which comes right up to the shore of the Indian Ocean, is an almost circular flounder that waits for its prey on the bottom of muddy shoals, stunning it with a series of quick electric discharges. For an instant it "produces" up to 200 volts at nearly ½ amps (equivalent to the amount of electricity needed to light a bright light bulb); this can even cause convulsions of the limbs in a human being, thus its common name.

After the discovery of the Leyden jar [a capacitive storage device for electricity, invented in Leyden, Holland, in 1745], the similarity of its spark discharges to the shocks these cramp-fish could bestow was first established. The following experiment done in 1773 became well-known: A torpedo fish was held wriggling between two nets stretched out in the air and dry (insulating) at the edges; a person only got a shock when he touched both the dorsal (top) and ventral (belly) sides. If the fish was touched with only one hand, the person felt almost nothing! Also, after a series of discharges over about one second, the following were very weak until the fish recuperated. What was more similar than the slow charging of the Leyden jar and its sudden discharge? As the electric activity of the electric catfish and the electric eel also became known, investigations into "animal electricity" became the fashion. Should we think of all animal activity as based on electricity? People seriously thought so at that time.

At the same time that Giovanni Battista Beccaria confirmed the permanent positive charge of the atmosphere, the attention-getting and overwhelming fiction was proclaimed by a confused public that all meteorological phenomena, right up to shooting stars, were electrically produced!

Today we know that the electricity (glavanic electricity) which we use daily plays almost no role in nature—with the interesting exceptions of lightning and electric fishes. The expressed phenomena that we are studying are those having an unmistakable relationship to experience; they allow us to live into natural forces and beings in a connected way, as we think about them in terms of what we perceive. In this sense these fish exhibit phenomena that characterize galvanic electricity. Beginning with the weightlessly swimming fishes, the hierarchy of complexity of bodily architecture points toward amphibians; then, in the struggle against gravity, the land animals exhibit higher sense organs, a larger anatomy and finally consciousness, but are less interwoven with their environment. This culminates in a personal countenance and biography in human beings; in this sequence, limbs appear. In contrast, the flounders and eels are limbless bottom-dwellers. Their fins (and also their eyes) are degenerate or atrophied. They sink down into gravity and into the dark depths; they show a kind of counter-impulse to the regular hierarchy of animals. Out of these murky depths, the capacity for electric discharge appears.

In his well-equipped home laboratory, the anatomy professor Luigi Galvani (1737–1798) of Bologna, previously a gynecologist, experimented into the reactions of frog-leg preparations, well-known for their sensitivity to electrical stimulation. As were many other people of the time, he was seeking to discover the soul of living things in the nerve connections. What took place in the summer of 1780 is reported by Galvani 11 years later when he finally undertook to publish his researches. (We'll see below why he waited so long.)

> This discovery occurred as follows: I dissected a frog and prepared it, as shown in figure Omega, Table 1, laying him unsuspecting on the table on which the electric machine stood, entirely separate from its conductors and moderately far away. However, as one of my auditors by chance brought the tip of his scalpel into contact with the inner side of the thigh-nerve of the aforementioned frog D, then the muscles of all the limbs were drawn together, as if seized by a mighty convulsion.
>
> Another of those present imagined to observe that it happened only at the times when the conductor gave forth a spark. I marveled at the newness of this situation and made myself wholly attentive to it, as I had originally intended something quite otherwise. I was overcome at that moment by the desire to discover and research these concealed things. I touched the knife point myself briefly on one and briefly on another thigh-nerve one at a time, while another auditor drew a spark. The phenomenon occurred in the same manner. Mighty convulsions took hold of the muscles of all the limbs, drawing them together, whenever the electric machine made a spark, as if the animal had been beset by a powerful tetanus.

This first discovery of Galvani's was nothing special and was explained somewhat by the so-called rebound effect, which occurs also in lifeless inorganic apparatus; so, these experiments were not the basis for his fame. The frog leg was only especially sensitive. But because the simple physical explanations of medical

Electricity

researchers—which he would have to apply here—did not explain these things completely satisfactorily, or else because he was not thoroughly familiar with such explanations, Galvani experimented further. Connected to long antennas, the frog legs were even able to predict the arrival of distant thunderstorms.

After numerous experiments of this type, one single observation in September 1786 (six years later) broke through into something fundamentally new: The frog musculature, freed from the skin which insulated it from the dry environment, was laid on an iron bar, in order to "take up atmospheric electricity into itself." It was hooked on an iron apparatus going into the spinal cord. Always, when the hook touched the iron rail, the thigh twitched. That happened—as Galvani soon determined—also in a closed room, at any time of day, with various metals for the support and the hook. The best results were obtained, however, if they were of two different metals. Thus, the source of an "animal electricity" independent of the atmosphere appeared to have been found: a kind of Leyden jar, formed from nerve and muscle, which even remained charged for some time after death—so Galvani thought. The external metal pieces merely act as conductors, to close the circuit of current.

The announcement by Galvani in 1791 presented something like this, was generally celebrated in the belief that now the soul of the human being and the animals was understandable as merely an electric effect, and electricity formed the basis for all phenomena of life. The sublime ideas still surviving in early German romanticism were dealt a death blow and displaced by the systematic research of Allesandro Volta, whose clear phenomenal thought was awarded a personal audience with Napoleon, at the same time as electric technology was founded.

But, in the same month of this year 1786, in a neighboring town in Italy, there occurred a totally different cultural-historical discovery. It was on September 27, in a botanical garden in Padua, that Goethe conceived the idea of active plant-formation: Within the sense-perceptible form hovered a super-sensible archetypal plant.[1] While the Goethean approach to understanding the plant opens to the human being a lifetime of maturing and transforming ideas about the plant and forms an ever-present realm of active fulfillment, the galvanic ideas opened up a realm of ever-present technical creations, which in its achievement had already lamed certain soul forces.

II. VOLTA'S PILE

But now, we leave aside historical events and follow up on the simplest experiments of Volta from the 1790s. The first of these is the very noticeable sour-bitter-sharp taste created by dissimilar metals in the mouth, when both are present at the same time. The taste is even stronger when they are touched in various ways (El1, El2, El3).

The creation of these sorts of phenomena goes in direct contrast to the way we produce electrostatic effects: Then we used lightweight thermal insulators under dry conditions. Now we utilize heavy, heat-conducting metals, always under damp or wet conditions. Instead of separating things under dry conditions, now objects must be brought together under wet conditions. How could we know that this sour-metallic taste is involved with something "electric" (as human senses cannot perceive anything of electricity)? From everyday experience, we know that something will taste differently when followed by tasting something else; so, the appearance of a distinctive taste when two materials are placed together on the tongue is not unexpected. It remains a mystery in El1 what is done by contacting the plates outside the mouth; it is a kind of spooky ritual, a magic formula. Then, to all appearances, nothing happens in the connection wire of experiment El3; only, we must follow the rule that it must be arranged in the ring-form and a break anywhere in it eliminates everything.

Overall, however, something else appears in contrast to other types of natural phenomena. Normally, throughout nature, phenomena develop in stages, and along with the daily and annual rhythms; they swell up and fade away. We can think of the weather, of the growth of plants, of heat and cold in nature. In addition, things occur in connection to a particular place: One process acts here, another there. In the sounds of natural things, we hear manifested their solidity, tension or size, i.e., when an apple falls, when a tree crashes down, or ice cracks on a pond in winter. The noise tells us something about the object and its location. With animals, sounds even reveal their inner state, what stimulates them or makes them suffer. And even in mechanics, from the splintering sound of wood or in the way it resists our arm, we sense the force that is directed to the active spot by a lever and fulcrum. But, what we call electricity shows itself entirely differently; connections in place and time are both torn apart. Electric effects are brought about with spooky, ritual-like procedures.

A part of the ritual, i.e., one of the conditions for occurrence of electricity, is the circuit of metal conductors and watery solutions, as shown especially in El3b and El3c. If, along the same circuit, everything is built in the same order of tongue touching copper and zinc, then the action is stronger. If all the zinc-copper pairs are reversed, the activity remains equally strong. If half are reversed, the activity almost disappears. The copper conductor in the wires is not varied at first—it only has to be present. The tongue functions as a so-called source of voltage (source of current) and simultaneously as the indicator. For the students, this doesn't have to be developed more closely.

With these simplest experiments, we can develop a concept of the abrupt/unexpected, cut off from nature, even nervous character of these phenomena. The systematic arrangement of these three experiments is more for the sake of the teacher's clarity. Students tire quickly of thinking things through for the sake of proof; one of the experiments may suffice, accompanied by the teacher's subsequently describing the sequence of phenomena, based on his first-hand experience. Now, the phenomena should be shifted out of the darkness of the mouth into clear glass beakers. First we demonstrate the "crown of cups" in a serial arrangement of voltaic elements, which illuminate the light bulb (El4); and finally the voltaic column, much celebrated in its day.

In 1800 in a letter to the Royal Society, the famous scientific society in England, Volta himself clearly described the construction of his column:

> The essence of this discovery, which encompasses nearly all others, is the preparation of an apparatus which in its action, i.e., through the discharge which it produces in the arm, etc., is similar to a Leyden jar or even more like a weakly charged electric battery, but which acts continuously …, which, in other words, possesses an inexhaustible charge.
>
> Yes, the apparatus of which I write, and which will doubtless cause great astonishment among you, is nothing other than the arrangement of a number of good conductors of differing types, which are arranged one after the other in a definite manner …
>
> I fashioned a dozen small round discs of copper, brass, or better, of silver, about the size of a toll-piece in diameter, e.g., like a small coin, and the same number of plates of tin, or even better, of zinc, of similar form and weight… In addition, I prepared a number of round discs of paper, leather, or similar porous material which is able to absorb a large quantity of water into itself. These discs, which I call the *moist plates*, are made somewhat smaller than the *metal plates* in order not to hang over their edges when they lie between them. When I have all

these parts ready, the metal ones well-cleaned and the non-metallic moistened with water or, better with saltwater and slightly dried, so that they will not drip, I only need to arrange them in a suitable manner.

And this arrangement is simple and easy: I place one of the metal plates horizontally on a table, e.g., a silver one, and place one of zinc on top, over which I lay a moist plate, followed by a second silver, then zinc plate, and again a moist plate ... I continue in the same manner building up more tiers until we have as high a column as will not fall over.

If it has been constructed to a level containing 20 or 30 tiers or pairs of metal plates, then it will be possible not only to charge a condensator (Leyden jar) through simple contact, but also to give one or more charges to fingers placed on both ends, which are repeated when one renews the contact ...

In order to receive a weak spasm, the finger must be moist, so that the skin—which otherwise will not conduct sufficiently—is well affected. In order to obtain yet stronger discharges, we must connect the base of the tower with a dish of water, into which we can place more fingers or even the whole hand, whilst one presses a metal plate onto the top of the column. We feel a prick in our fingers then, when we contact only the third or fourth pair of plates; if we now touch the fifth or sixth, it is interesting to note how the spasms show a stepwise increase in intensity...

One of these observations—that the column is inexhaustible—is overstated, because the electric light and heat which we produce are hardly produced from nothing. The glowing electric bulb (El4) offers one application—where is the consumption supported? If we allow such a bulb to burn for a while, then one plate very clearly undergoes a change: The zinc is eaten away and white zinc salts mingled with zinc hydroxide cover the plates. The voltaic column must be washed free of this residue, an inconvenience which the other source of current doesn't have; the "crown of cups" is otherwise impractical, since it is non-transportable, cannot be maintained in activity, and requires a space many times larger than the Volta column.

In Volta's time, people attempted to make the mystery of electricity perceptible by charging their bodies in all sorts of ways. Thus, Volta reports:

Just a word about hearing: This sense which I had earlier attempted to stimulate without success using a pair of metal plates—even though I used the choicest metals, namely silver or gold with zinc—I have now been able to stimulate with the new apparatus of 30 or 40 plate pairs. I inserted two metal rods with round ends into the ears, and then connected these with the apparatus. At the moment of connection, I got a shock in the head, and somewhat later, while the conductor healed, I heard a tone, or more precisely, a rushing in my ears, which remains very hard to describe. It was a kind of knocking or rustling, as if a tough material was being torn to threads. This noise persisted without becoming louder or softer, as long as the current was active. The unpleasant feeling I took to signify a danger, as it brought about a convulsion of the brain. Therefore, I did not repeat this experiment.

While the voltaic column has almost vanished from electric technology, the crown of cups was an inspiration for all kinds of dry cells and batteries. These are used today by the millions, from a Geiger counter to a cardiac pacemaker. The simplest battery consisted of a copper-zinc elements in a copper salt solution, widely used in the telegraph stations of the last century (the Daniell cell type).

Later, the zinc-manganese dioxide element was invented (Leclanche cell) and became the basis for the modern dry cell used in most flashlights. For more details, see the Accessory Topic on Galvanic Cells.

III. VOLTAGE AND CURRENT

No matter what we try to find out about the relation of electricity and outer conditions, we can never get to the essence of electricity; it remains concealed. Still, we can list the conditions for the effects it creates in the human body (spasms, burning) or in a light bulb (heat, light):

1. There must be two dissimilar metals: one as noble (precious) and the other as base as possible, not touching but immersed in a salt, acid or alkaline solution which makes the internal portion of the "circuit."

2. An external connection must exist (be made) from one to the other electrode: via a metal wire or some sort of circuit.

3. The body or the light bulb can be inserted into this external pathway, having contact in two places. If we want to produce glowing or light, a piece of thin, conductive wire must lie across the contacts.

The most essential condition is the first. The second allows a great variety of numbers and kinds of materials. In contrast, the first is very specific and plainly is the most important. It can be demonstrated (as was done in Volta's time) that the ends (terminals) of a voltaic column or pile show electric field effects. Even when no current is active, e.g., the leaves of an attached gold-leaf electroscope will spread apart. *In addition, the column ends act like charged poles in electrostatics.* But, with a current-conducting circuit, the ends show no charge—they are ordinary wire.

These delicate repulsive effects are familiar to us from our work with frictional electricity and paper strips, but are usually only observable in a technical laboratory at the terminals, have led to imagining the voltaic column (even without a current) as having an electrical condition—namely an "electrical potential." It is a source of electricity, i.e., a kind of point-of-departure for everything that happens. Electric tension or charge (technically, potential difference) is a *static condition* between two objects (electrostatics) or between two parts of one object (galvanic electricity), i.e., the potential between the poles of a flashlight battery, which contains electric phenomena in readiness. Electric potential means we can later produce phenomena that are presently potential, i.e., have not yet occurred. Thus we can buy tension in the form of a flashlight battery, which is somehow "charged" full of voltage (electrical potential), and the battery serves as a supply. The greater this voltage (tension), the more charged the battery is, and the more one can produce with it. With 1000 volts, the mere proximity of the poles of a spark gap is enough to cause a resounding discharge. Thus, high voltage requires a great separation (in air) or thick insulation. In electrostatics, when we set our charged tin-canisters on glass bottles or on Styrofoam plates, they had an insulation distance of about 1 cm. In galvanic electricity, very little insulation is needed; the insulation actually serves merely to prevent a short circuit—the current must be insulated against connecting back upon itself.

When, as in experiments El4 and El5, we make a circuit connecting the poles of the charge source, then out of calm potential, a two-fold process arises: Here, the wire in the bulb glows; there, the zinc plates are eaten away, leaving zinc salts (and simultaneously at the copper plates—although barely evident in the phenomena we perceive—hydrogen bubbles out of the solution and a weak nitric acid solution arises). There is a relationship between the development of heat and light with the consumption of the metal electrode and the transformations occurring in solution. Whether the conductor is a mile long, runs uphill or downhill, deep into the earth or up into a balloon, this relationship holds. This relationship between light (glowing) and consumption of zinc at the cathode is termed "current." Of course, nothing "flows" in the wire. If we have more bulbs, then after closing the circuit they all burn simultaneously, without any delay, rather than the one nearer the electrode or the switch glowing sooner. With electric current there is no "front" which passes along the circuit, nor is there a direction of motion along which it streams.[2] Current is not something objective in the world available to us, but it is an idea of relationships existing in the active circuit In any case, the current is a condition existing in the conductors and the intervening fluids, not a process where something "runs." Therefore, we should say current "exists"—rather than current "flows." To produce current, then, is to say that we permit a relationship to exist between phenomena of consumption or expenditure (metal) and yield (glowing bulb), and does not mean we set into motion a quasi-material something in the wire. In general, it means that outside the wire a magnetic field is created (see grade 8); and inside the wire, a gentle warming occurs, corresponding to the cross-section of the wire; it is small in large wires but never entirely disappears. So, within the wire itself (which is only included as a necessary condition), nothing at all occurs as the result of current other than a germinal beginning of something there, namely the slight warming (warmth as the beginning of anything that happens).

IV. CURRENT STRENGTH AND SHORT CIRCUITS

This current, carefully conceived in this sense, can be more or less powerful. It is not necessary to abolish the conventional idea of current strength. Large current means a rapid consumption of zinc, i.e., one gram of zinc is used up in a shorter time. Concurrently, light bulbs produce much more heat and brighter light. A process having a magnitude in all parts such that in 1 hour approximately 1 gram (1.22 g precisely) of the zinc plate is consumed (and therefore appears as zinc salt in the moist layer), by definition produces a unit current = 1 ampere.

To enable the electrodes to produce a greater rate of consumption of zinc, i.e., a larger current, they must provide a greater surface area in solution, or else must be porous. To be able to force through a larger current against resistance (thin wire, many bulbs arranged serially on the wire), we need higher voltage, i.e., more plate pairs, arranged as above.

While voltage is a capacity or potential, current is an event, usually in two places. It can be limited either by the rate of consumption of zinc possible in the battery or by the resistance, i.e., by the conductance of the bulb filament in relation to the voltage available. Current is the actuality and voltage is the potential for penetrating conductors.

With the concept of current we can now more closely investigate storage batteries (lead wet cells), overload protection (short circuits) and light bulbs more closely. The dissimilar metals needed in the automobile battery are lead on the one side and brown-black lead peroxide on the other, even if only as a thin layer on the lead plate (El6). For larger current and output, flat plates do not have enough surface area, so commercial storage batteries are constructed with porous plates (El7).

The automobile storage battery can produce 20 or even 50 amperes. Therefore, a conductor draped over the arm gets hot (El7). The hottest place is where the conductor is thinnest: At the frayed strands at the end of a wire, the contact surface is often so small that the strands melt, glow, and even vaporize, explosively throwing out sparks to all sides. If a circuit consists of thick wire and there is too little resistance in it in the form of a thin, poor conductor (a light bulb inserted into the circuit), then it will draw too great a current, and the wire can get overheated and could start a fire. The experiment (El7) shows the appearance of an overly strong current and the sparks characteristic of it. Such an intense heating is connected with an equally intense consumption at one lead plate in the battery. The plates must be correspondingly large; a clear relationship exists between conductor thickness and plate size (area). A thick wire or a thick cable (with many strands connected into the circuit) produces a great heating, requiring equally large plates. From where does this excessive current arise? The construction of a circuit of only low resistance conductors calls it forth. The battery makes it possible, certainly, and permits itself to be useful for a smaller current.

This can be the next experiment with a suspended resistance wire (El8). Such a wire is thinner and a poorer conductor than the copper connection cables. The entire heating occurs in it, and it acts as the limit for the "metering out of the phenomena," the current. If we shorten the length of thin wire, that piece gets hotter and the current is stronger. The resistance of any wire is proportional to its length. If we shorten it further, the heating and the associated current will increase to the melting point, and the wire will finally melt through. Copper wire, which melts through already with a longer length, shows its greater conductance; copper is a material of little resistance, therefore good for cables.

Initially, current, voltage, and resistance can be understood by the students only in a qualitative-experiential way and only in their experimental significance. There it can remain in grade seven. However, the teacher ought to be able to use these qualitative concepts and recognize the objective correspondences and magnitudes implicit in them without having to do precise calculations with formulas—also to be prepared in the moment to give answers to questions in class.

Electricity

V. VOLTAGE AND RESISTANCE

What is voltage (electric potential)? Voltage is a condition of readiness of electrical phenomena; a condition capable of completing the circuit [with a spark] without anything further being done, for example, when we approach a rubbed foil. Scientists speak of the availability of charge in objects, so generally we imagine it can be produced by rubbing, separating or dunking metal plates, etc. As an objective phenomenon, however, voltage (potential) is nothing. If electric phenomena are present, it is not the slumbering voltage (potential) but the various movements or heating (effects) that we perceive; thus, we can't understand the situation with the concept of *voltage* alone. In the ideal, i.e., largely approaching the reality, voltage (electric tension) is, strictly speaking, the pure possibility of electric events; it is an event-preparing condition, a state of waiting.

This inclination toward an event (tension) can be more or less powerful; with high tension, sparks can jump a distance of yards. A quantitative measure of this tension, i.e., the numerical value in volts, specifies the capacity for penetrating the environment compared to the penetrating capacity of a 1-volt source. In contrast to electrostatics (grade 6), our galvanic experiments with wet-cell metal plates (grade 7) produce only low voltages: 1–2 volts. They cannot produce so-called "high-potential sparks," which cross the gap before contact. But how can we increase the voltage from our copper and zinc plate cells? By inserting more cells (El3 and El4,

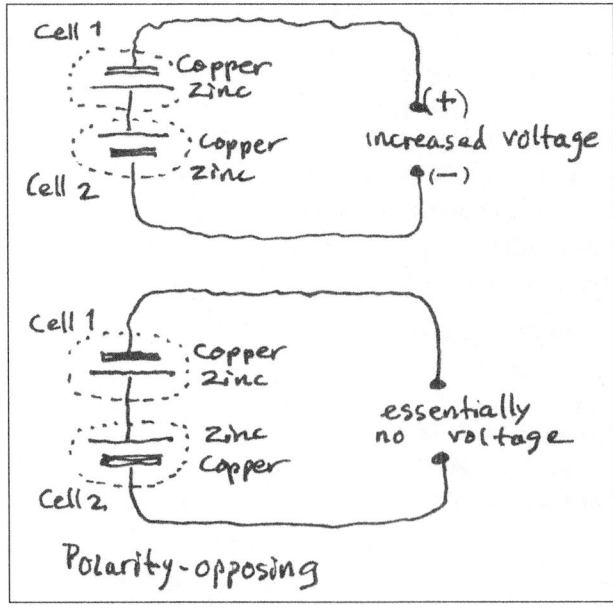

Figure 19. Connecting cells to increase voltage

crown of cups). For the voltage alone, it is unimportant how large the plates are—we could use the metal strips of El3. The order of insertion is decisive.

How does this arrangement increase the voltage? Electric tension exists always between two poles. In Figure 19 top, one pole of a cell is connected to an opposite pole, e.g., plus, and the other to its opposite, minus. The copper electrode here is positive.[3] In contrast, the circuit with both zinc plates connected together (Figure 19 bottom) makes both copper plates positive to the same degree. In relation to the common reference point (the connected zinc plates), the copper plates have the same potential, so zero voltage (potential difference) is produced between these two plates. Since voltage is characteristic of a battery, such a cell is worthless. A battery with a greater voltage, i.e., a larger ability to produce events, is also able to produce sufficiently strong electric phenomena, even in a long, thin wire that is a poor conductor (e.g., in a light bulb); we call this current. A battery with greater voltage increases the power of the events: the bulb burns through. The battery's voltage and current must be matched to the circuit's resistance.

Resistance in a circuit depends on the material of the wire. Everything plays a role, which is directly a part of the circle of conductors (i.e., all components whose removal would eliminate the current-phenomena). The resistance of any particular piece of wire (or any portion of the circuit) depends on three things: length, thickness (cross-section), and material (composition). In our experiment El8 with the glowing wire, the power cord (to our powerstat) can be made longer or shorter with hardly any effect on the heating of the thin wire. But if the thin wire is shortened, it begins to glow much sooner and more brightly; it is that part of the circuit which determines the current (the scale of the overall phenomena). It is the limiting resistance. The resistance of the power cord must be assessed as much less, since it has little influence in contrast to the thin wire segment; so, we need not actually consider it further in this connection. Experiment El8 shows that the resistance is smaller with a shorter, thin-wire segment since the current (and the heating) are greater. Now we imagine two such wires stretched parallel. The same warming occurs in each as earlier in one; the battery is consumed twice as fast, and overall there exists twice the current.

Experience also shows that without altering the overall current, the neighboring wires will be drawn together and can even melt into one. Then we have one wire but with double the cross-section. It regulates the whole circuit to twice the current as compared to that acting in one of the initial (thinner) wires. Double the thickness has half the resistance. In general: Electrical resistance increases with length and decreases with cross-section.

Electricity

To investigate the properties of these materials, instead of the iron florist's wire, we could insert wires of different composition, but with the same length and cross-section. *The material which heats the most conducts the best.* The exact current strength can be determined from the speed of consumption of the battery plates. Naturally, in such an experiment, we attempt to use cables to the battery with a much smaller resistance than the wire in question. How is this achieved? They must be thicker and made of a better conductor. In our experiment, they were twice the diameter and (most important) made of copper, which conducts 8 times better than iron. This principle forms the basis for all household appliances and for the whole electric network: The conductors must have a resistance much less than the object being used; i.e., in comparison to the thin and long, coiled-up filament of a light bulb. And, just there where the greatest resistance lies, the greatest heating occurs. Thus, the thick plug contact castings, or the support wires in the light bulb, each have a much smaller resistance than the helical filament.

The size of the battery, specifically the area of the plates and not the number of cells, plays a decisive role. Initially, this can be summarized in the rule: The current in a resistance circuit increases with the size of the associated voltage and decreases with the size of the resistance. If we make the resistance smaller, the current grows larger and larger, up to a limit, as the battery cannot power any arbitrarily large current. Its voltage potential fails. For example, a car cannot be started with a 12v transistor radio battery, even though the automobile battery also is 12v. However, its porous lead plates have almost a thousand-fold greater surface area than the zinc wall of a tiny dry cell. The large cell plates permit a rate of consumption 10 times greater, since the eating-away occurs at the surface of the plates. The maximum current occurs when the circuit has the smallest resistance, and even no longer includes an external wiring but is limited only by the plates themselves. From the size of the cells, we can approximate the maximum current, and from the number of cells, we can know the voltage precisely.

VI. THE EXPERIENCE OF ELECTRICITY

What opinion, what feeling should the students gain toward electricity? We are tempted to respond quickly: Just present the facts and leave them free. But that is not what should happen, at least not with youngsters. In order to teach a subject, a connection must be created between what we present and a deeper experience in the students, with their inner life. Otherwise, the developing kernel of the personality is weakened. We have no illusions that cold manipulation, glorying in our human knowledge of experimental effects, or egoistic applications of these so-called supersensible natural forces have no feeling-connection with the world. An approach that claims to be value-free and only teaches facts related to one subject, in fact, does have a feeling connection to the world, but a world unconnected, separate and cut off. Teaching steeped in such a worldview makes a deep impression, even if it is considered objective education.

When pursuing a subject, even if the feeling-orientation is not expressed, it is certainly good to make this feeling-orientation consciously formed for the students and their parents, to touch the spiritual in the world that arises in them. That is very individual for each teacher and each class. Nevertheless, the orientation we have used here can be summarized in three aphorisms, which are:

(1) *How electricity pierces time and space*, or, more precisely, the perceptible transformations in location, the changes in temporal sequence that we experience, by which our mind defines our world, cannot be properly built-up with electricity. Without warning, electricity severs the tapestry of coherent natural images. We compare the way it appears to nervousness, although the words are not what matters.

(2) *How every circuit bears its tension isolation*. The plus terminal of a battery can only produce current phenomena with the minus terminal of the same battery. Thus, a wire segment can simultaneously be part of two circuits without appreciably altering the working of either. For small currents, the circuits influence each other only via the heating of the common conductor (which slightly increases the resistance) and the related expansion. In the following circuit layout, the cross-effect is barely noticeable:

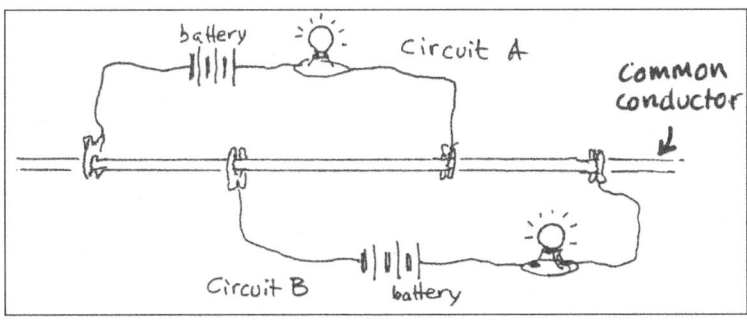

Figure 20. No cross-effect with electricity

Every circuit acts as if aware only of itself, of its own image of the world; it is closed up in itself. Things going on adjacent in the surroundings have no effect; it is oriented only to its own properties. This is the reason we can carry circuits around inside us, e.g., in a hearing aid. For acoustics, in contrast, the tone depended not

Electricity

only on the string, but on the entire instrument, even the surrounding room. Brightness is linked up with the most distant places. Warmth changes according to the warmth of the surroundings. But the circuit, with its non-natural conductors or insulators (in contrast to tissue, leather or wood), inserts itself into the environment as something autonomous, foreign.

While the first aphorism is oriented more to thoughts about electricity, and the second more to our feelings about it, the third points out the (3) *limitless possibilities of what we can do*. If the students fully understood the concepts of voltage, current, and resistance discussed above, then they will be able to rationally produce and manage any sort of electrical effect with understanding and without needing any sort of deeper feelings about its connections to nature. Mentally rearranging the external connection of inexpensive and weak batteries can, without any other change, make a stronger power source; the shifting of a mere powerless wire, to rearrange the connection, is sufficient to increase the voltage (series wiring) or the current (parallel wiring). Insertion and reversal of circuit elements can produce the greatest effects without much effort—i.e., an abstract or magical result: a switch.

Goethe's aphorism (see his Color Theory) is either the most important or says it all: The meaningful ethical-moral should be given our attention. These sorts of things ought to be *felt* by the students and be *actively considered* in discussions. A bit clearer evaluation of this could be sought—perhaps after a year's sleep—then, in the 8th grade, in such a way that we pursue the consequences which our awesome electrical civilization has created for the life of the individual, for the social organism, even for the whole earth, so as to awaken the ethical-moral impulses necessary to counterbalance them.

Experiments in Electricity

EL 1 ELECTRIC TASTE

Have one student examine the taste of single zinc and copper plates; they taste like bare metal but can hardly be differentiated. Nothing is altered if they are both touched to the tongue simultaneously. However, when the outside ends of the strips are brought into contact, a sour, somewhat prickly taste appears, more strongly with the zinc strip.

EL 2 DISTANT TASTE

Attach connection wires to each strip with alligator clamps. When the connection is made with these wires, the same prickly taste appears immediately. We can even join them behind the students' backs. They will notice

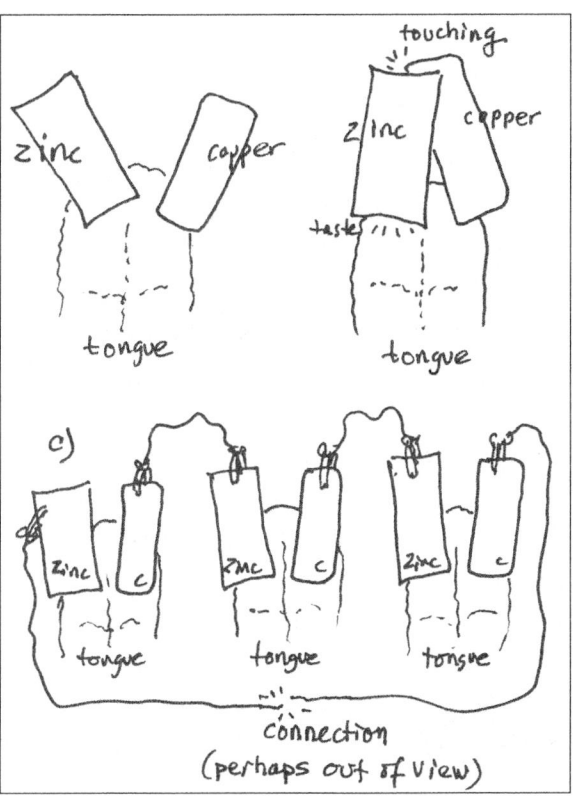

Figure 21. Tasting electricity

immediately when the connection is made; even if the wires are run to the far end of the classroom, they can correctly tell. (We have them show when the plates are connected by stamping their foot.)

EL 3 INCREASING ELECTRIC TASTE

Using two connector wires and a copper and zinc strip for each, we can form a human "battery of tongue cells" with several students by connecting in serial the zinc and copper strips they taste (Figure 21c).

a) The students indicate (by stamping) when the electric taste appears. The instant the circuit is completed, they will all stamp in unison (aside from the phlegmatic ones).

b) If just one student removes either metal strip from his mouth (or one connection comes apart), everyone will taste nothing.

c) Instead of connecting the wires to each other, join the clamps via: (1) a wooden strip [no taste]; (2) a tin can, scissors, anvil, iron pipe or clean-stripped heating duct [electric taste].

EL 4 CROWN OF CUPS

The "Couronne des Tasses" (crown of cups) is a series of beakers, arranged in a circle like a crown. Use 3 or 4 cells, each with a pair of copper and zinc plates, set up in beakers filed with saturated salt solution (table salt in dilution water), all in the same sequence: zinc, then copper. When the wires from El3 are connected to a small bulb, it glows! [Try a 2v, 0.06a bicycle bulb or else,

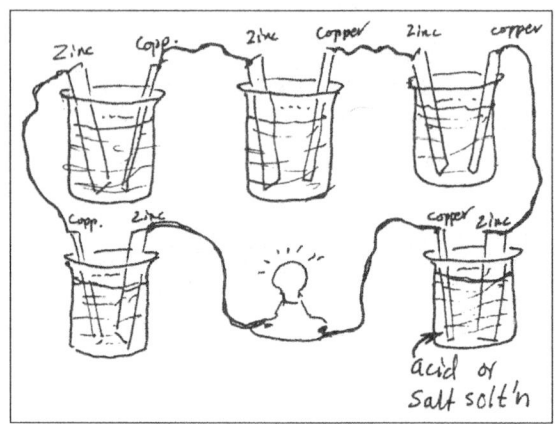

Figure 22. Crown of cups

albeit hi-tech, a 2v LED.] A larger (1.5v, 0.1a) bulb will still glow, albeit weakly, with one plate pair (voltaic cell), which produces about 1 volt with copper and zinc.

Instead of connecting the wires to a bulb, they can be touched to the tongue The strength of the taste corresponds to the number of cells (and brightness of the bulb's light).

EL 5 VOLTAIC PILE OR COLUMN

Build up a pile of 3 to 5 pairs of plates with interspersed with paper discs of filter paper (or blotter paper or newsprint, if necessary) saturated with vinegar or salt solution. Copper follows zinc. One plate contact is dry, the next via a moist plate; the zinc plates and even the copper should be polished on the dry contact sides to provide good contact. Each pair should produce at least 0.7v. The solution should not drip down across the plates beneath; the alligator clamps on the end leads should not touch

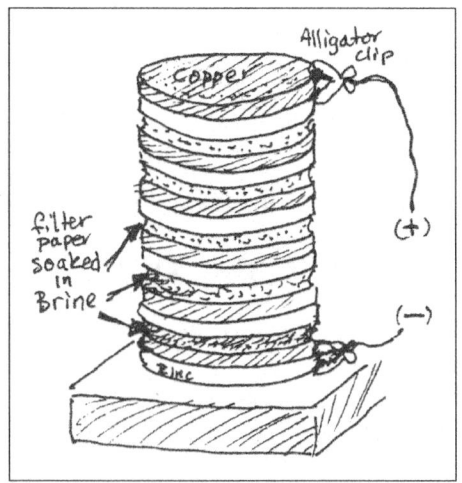

Figure 23. Voltaic Pile or Column

the next plate. Often it is necessary to clamp down on the whole pile of plates. The demonstration is best done in silence—without prior explanation. Again, the bulb (from El4) glows and the leads taste "electric" on the tongue, showing electric current has been produced. [Clearly, test out this set-up before class.]

EL6 ACCUMULATOR OR STORAGE BATTERY

Two pieces of lead foil are crimped by making cross-hatched creases with a knife, bent around the wall of a beaker, then placed in a beaker with 20–30% sulfuric acid solution. [With CONSTANT stirring, SLOWLY dissolve 66ml of stock concentrated sulfuric acid into 133ml of *cold water* in the beaker—NOT the other way around—to make 200 ml of battery acid solution. Wear safety goggles, and take care not to spill! This can burn holes in clothing. Dispose later by flushing down the sink with copious amounts of water.] Show how a bulb wired in (as in Figure 22) shows nothing, for this is not yet a productive voltaic cell. Allow the plates to "get charged." [Using a variable DC power pack or medium power battery charger, apply a DC current for 1 hour or until noticeable gas bubble production occurs. Take care the plates do not touch during charging. Separate them with a spacer: a piece of polyethylene mesh or fragment of a milk-jug punched full of holes.] We see the plate connected to the positive lead develops a brown-black coating. We have made a storage cell: Now, the small bulb (1.5v, 0.1a) wired in will glow when connected.

EL 7 CURRENT SPARKS

To show the heating effect of current, prepare a 3-meter piece of multistrand (house) wire by stripping about 30 cm of insulation off one end. Screw on large alligator clip connectors on the other end. From an insulated clamp on a lab stand, hang a second one so the stripped part is suspended over a car battery. [Used but serviceable, one can have the top removed, exposing the cell connections. Take CARE, as noted above, with the battery acid.] Clip the dangling end onto the negative terminal. Drape the other cable three times around the bare arm of a student wearing safety goggles. Hand the stripped end to a second student (also equipped with safety goggles) who holds it ready. The teacher now fastens the other end, via the clamp, to the 4v terminal (second internal cell-to-cell connector in the battery).

The second student now has the job of stroking his bare end against the bare wire suspended over the battery. The teacher increases the voltage stepwise [by clamping onto the next positive cell-connector, and then finally onto the main

Experiments in Electricity

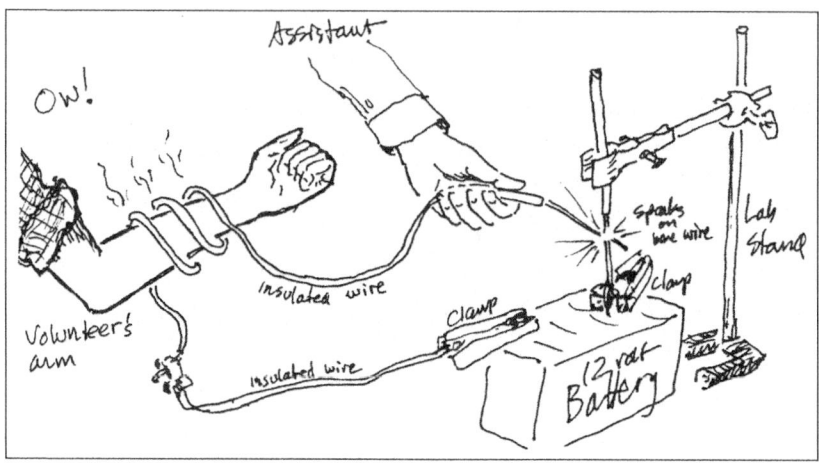

Figure 24. Experiencing current heating

positive terminal]. Eventually we get a powerful shower of greenish sparks, glowing tiny blobs of metal falling off, and a very noticeable heating of the coiled wire.

Different colors are produced if we use other metals. A steel rod (not chrome-plated) is placed horizontally on the lab table and connected with ordinary cable to the negative terminal of the battery. An iron wire, grasped with a potholder is connected to the positive terminal and then tapped onto the rod: Orange-colored sparks appear, characteristic of iron. Usually, with dry hands it is possible to touch the wire that made such spectacular sparks without any danger or noticing anything; also, the whole top of the storage battery (if clean of acid) can be touched without shock—even though it is capable of melting thick wires! [*Current produces heat and magnetic effects; voltage has the potential to spark across non-conductors. Although capable of 30–50 amperes, our battery produces only 12 volts, which is far too small to carry across dry skin; thus, no danger. Also, try experimenting with the various metal rods used in Experiment He2. – Transl.*]

EL 8 RESISTANCE AND CURRENT

A piece of iron wire (florist's wire) about 1m long and 0.5mm diameter is suspended 0.5m above the lab table using two lab stands and insulated clamps. Using somewhat short connecting cables, we apply about 2–3 volts DC from a variable power supply (varistat): a student reports that the wire is getting warm; a folded paper flag in the middle of the wire drops, showing the wire is sagging (heat expansion). Increase the voltage to 12 volts; then clip on to ever shorter segments of the wire with the alligator clip on the connection cable: soon the paper flag smokes, eventually falls off burning as the wire begins to glow; with further power, the wire melts through and glowing pearls shower down (most don't burn as they roll away); the melting wire tears under its own weight. A piece of copper wire of the same thickness melts through already while the connected segment is much longer. It is a much better conductor and draws a larger current (heat) sooner.

ACCESSORY TOPICS

- Wiring circuits parallel (increases amperage from electric cells) and serial (increases voltage)
- Types of batteries:
  a) zinc-carbon or Leclanche-type dry cells (damp cell)
  b) lead-lead oxide automobile storage battery
  c) nickel-cadmium rechargeable cells

Grade 7 Magnetism

INTRODUCTION

In the 6th grade, we presented magnetism mainly as a directionality within the earth's environment. We produced magnetism from terrestrial magnetism and studied its effects. Now, we want to direct our attention first to antique and then to modern, manmade magnets. We turn from the orientation exhibited by the whole earth, from sailing and the oceans, to explore hidden poles in magnetized pieces of metal. A main theme that comes out of experiments about proximity to strongly magnetized pieces, especially with iron, is the augmentation of the magnetism by direct contact. Everything in the vicinity made of iron is "influenced." That is a deep experience, almost a "manifest secret" (Goethe) about magnetism.

The Romans already knew of lodestone (magnetite). Pliny the Elder, in his great treatise on history writes:

> What is more inert than unfeeling stone? Yet, see how Nature gave it feelings and hands. What resists more than hard iron? And, yet it will yield and be obedient; for, it can be drawn by lodestone. And, this otherwise compelling metal, here meekly follows an unknown substance; and if it comes near, stands still in fetters, lying in its 'arms.'
> – Historium Naturum, Lib. XXXVI, c.16)

Antique magnets were made of lodestone pieces fitted with iron plates; the natural stone piece was "suited up in armor." When a "yoke" was hung from this "suit," it usually should support 30 or 40 times its weight.

By mounting it in an armature, the part of the lodestone that tends toward

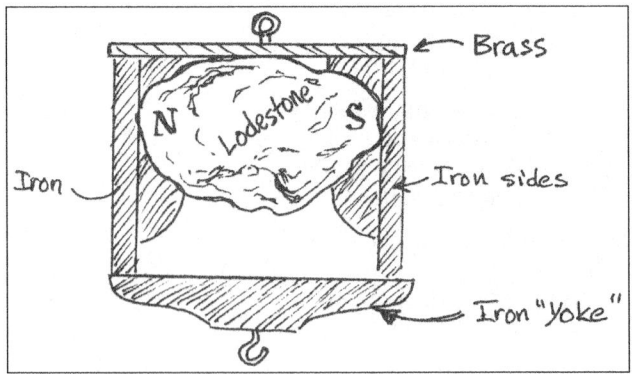

Figure 25. Lodestone mounted in armature

north is linked to the south-pointing portion via the yoke; thereby we increase the carrying-power significantly, at least a hundredfold. Usually small lodestone fragments can support more load in relation to their own weight because the magnetic force in them has a relatively smaller surface area to work through.

Goethe had investigated such an armored magnet in his childhood, as he described in his autobiography:

> An armored lodestone, very attractively sewn up in scarlet cloth, one day also had to experience the influence of the delight for exploration. For, this mysterious attractive force, which not only was exerted upon a small iron rod pressed against it, but was moreover of such a sort that it was augmented and daily able to bear a greater weight, this mysterious virtue [influence] had overcome me to such a degree that for long periods I fell into naked astonishment at its activity. At last, I hoped to gain a more precise elucidation by freeing it from its outer sheath. This was done, though I became none the wiser, for the naked armature taught me nothing more. I put it down and kept within my hands the naked stone, which I never tired of using to make the most varied experiments with filings and nearby needles, from which—aside from manifold experiences—my youthful spirit drew no further benefit. I did not know how to put the whole contrivance back together, the parts were strewn about, and so I lost the eminent phenomenon together with the apparatus.[1]

How can we lead the students into this realm? Initially, we show the ends or poles of two large, free-standing compass needles. Certainly, just as north-pointing ends placed underneath each other will repel, a north-seeking near a south-seeking end will mutually attract (Mag1). This demonstrates the well-known rule: *Like-named ends repel; unlike ends attract.* In light of this fact, the north magnetic pole of the earth must be described as magnetically south, since it attracts the north end of a compass needle.[2] If we move a compass needle around a bar magnet, we see how the attraction seems focused at the ends; the exact middle of the bar seems indifferent (Mag2). The center has much less capacity to make iron filings stick (see also Mag5). So we term the ends *poles* and think of them as if the entire magnetic activity proceeded from them alone. The location of these poles is neither at the end faces nor points just under their surfaces, even though we saw the needle point directly to the end surfaces.

Nevertheless, many magnetic phenomena can be explained with this idea of localized, point-like poles. If a bar magnet with two poles, one red type (north) and the other blue (south), is cut—without too much vibration—into two halves revealing a previously concealed cut surface, then each half will also have two poles, each as strong as the originals! (Mag3) Thus, the poles cannot be the fundamental cause of the phenomena, since they appear and vanish at will. What is constantly present is the orientation of the magnetization. Where this passes out of the iron into the air, naturally, is the boundary of the metal bar; this is also the limit of needle movements outside—so we speak of a permanent pole.

The attraction of iron, the most familiar phenomenon, is studied too. How can we explain the often-observed attraction of yet unmagnetized iron filings? At first, these are completely nonmagnetic. But, in the vicinity of a magnet, they become magnetized—clearly in the same direction. So, around the "red" (north) pole of the primary magnet, a "blue" (south) pole arises secondarily in each filing—and thereby attraction for the primary magnet. But, at the other end of each of these secondary magnetized filings, there must be a complementary "red" (north) pole. And, any other iron filing in the vicinity of the primary magnet, and close to the first magnetized filing, will be secondarily magnetized in the same polarity as the first filing. Thus, it will be repelled by the first filing (as Mag4a shows), at the same time it is attracted to the primary magnet. In these experiments, the primary magnet acts like the whole earth did for our compass work last year: Everything in the vicinity becomes magnetized. However, the earth does not act attractively in any locales (despite legends of a magnetic mountain), only directionally. Clearly, this directionality will be focused and concentrated by large ore masses of iron, i.e., deflected and increased in intensity. This will manifest in deflection of the compass (aberration).

The magnetic field is one of the most enchanting and beautifully-formed phenomena of magnetism. How are those line drawings produced? Initially, the parallel curved lines are obtained when we join the individual orientations of a compass needle at each position, as we move it along arbitrary paths from pole to pole around the bar magnet In this way, the field lines can be derived precisely from the magnetic orientation, but this would require days. Instead, myriad iron filings will show it in an instant (Mag5). To explain this, we recall the section above: The iron grains are magnetized. If they lie transverse to the field, they are pulled into alignment, as was the compass needle in Mag2. Also, from the clarity of the resulting pattern, we can read off the magnetic field strength. It decreases with distance. Clearly, the pattern arises due to the light weight of the filings. In free space, the pattern of force that we describe by the term "field" is not limited to the locus of thin lines; rather, it acts upon a magnetic compass needle everywhere, homogeneously and continuously. The field is a reality without anything being there, a reality in an idea (a coherent summary) of all the directions and orientations given by all the various compass positions. At the same time, it is the magnetization potential for particles subsequently strewn in its vicinity. Thus, we have three steps:

1. There is nothing around a magnet. However, when a compass needle is brought near, it is deflected.
2. If we make a sketch of these possible needle orientations in the entire space about a bar magnet, we thereby describe the "field."
3. If we sprinkle iron filings there, then this potential for orientation will be realized and form lines at arbitrary distances, as thick as the particles we have sprinkled.

With a bar magnet, the filings show clearly that the magnetization does not emanate only from two so-called poles, although they are concentrated there. The magnetic field of an artificial permanent magnet, revealed by the iron filings, is of necessity overlaid on the earth's magnetic field; i.e., the magnetic needle takes up an average position for both, depending on the relative strength of each. (This is significant in Mag2.)

Shielding the effect of a magnetic influence, or if you prefer, of the field, can not be achieved with paper, wood, glass or brass, nor with aluminum or water. But, a steel plate oriented diagonally or, even better, an enclosing steel sphere, will permanently shield the effect of a magnet (Mag6). A steel rod may focus the magnetic field within itself if it is placed between the "sensor" compass and the influencing bar magnet. It will take on an induced magnetism itself, and thereby the magnetic needle reacts. The diagonally oriented steel plate, in contrast, becomes a magnetic boundary; it acts like the middle of the magnet, indeterminate, so the influence of the magnet is shielded. If we approach too closely to the shielding plate, then the phenomena of proximity will be seen. [The needle will be attracted to the metal plate simply because it is metal.] Because the field will pass through non-shielding materials, we can do experiments such as the concealed magnet experiment (Mag7).

Permanent magnetism requires special materials. We can produce a hard or permanent magnet using materials that retain a permanently strong magnetism, even after removal of the producing field, in contrast to the so-called weak magnetizable materials. Strictly speaking, only the metals iron, and to a lesser extent cobalt and nickel and alloys of these (i.e., 5% nickel-steel), can be magnetized. Such materials are called "ferromagnetic," named for iron (*ferrum* in Latin). Magnetic pyrites and magnetite (crystalline iron oxide) are weakly magnetic. Iron and steel with a greater hardness are more difficult to magnetize but retain the magnetism, even resisting the demagnetizing effect of mechanical vibration. Steel that has been freshly quenched (see grade 8 chemistry) or chrome or tungsten-steel is the best for permanent magnets. The most powerful is the alloy discovered in 1950: "alnico," an alloy of 43% iron, 33% cobalt, 18% nickel and 6% aluminum. Today, most powerful demonstration magnets are made from this alloy. It is produced—as most magnets are nowadays—by the unsurpassed electromagnet (see grade 8, Mag2 & Mag3). Further topics, not as necessary or impressive but useful for exercising our thinking about magnetism are found in the Accessory Topics supplement.

Experiments in Magnetism

MAG 1 ATTRACTION & REPULSION

While two large compass needles (demonstration compasses) on stands are pendulating toward the north, we take one off its support and hold the "red" (north-seeking) end underneath the "red" (north-seeking) point of the other, which now responds. [Like-named ends repel, different-named ones attract.] If available, we show two ring magnets (floating magnets); they can even be rolled on the table and will jump in an instant onto each other. In our hand, they exhibit an almost tangible repulsion for one placement [the ring has its poles on the broad disc surface, not on the smaller edges like a bar magnet]. We can also show how these ring magnets will successively influence a series of compass needles, making them swing from SE to NW as they pass overhead.

MAG 2 POLES

With a small compass needle, we explore the vicinity of a large, permanent bar magnet, positioned on a wooden block (Figure 26a). Precisely in the middle, the needle lies parallel to the bar; nearer to either end, it points toward that end. It is possible to get confused in this exploration, since the "red" (north-seeking) end of the needle will be attracted to anything ferrous if it comes too close—even the "red" (north-seeking) end of the bar magnet! When near the bar, the needle can be weakly deflected for this reason. Also, the needle will be attracted to any nonmagnetized piece of iron (Mag4). The solution is to maintain a certain distance, usually twice the length of the compass needle from the bar.

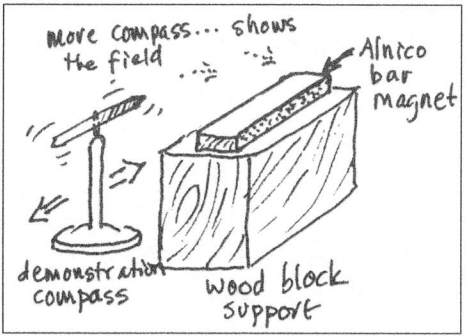

Figure 26a. Magnetic field

MAG 3 DIRECTION

Prior to the class, we set up two short, thick *magnetized* rods of construction steel, laying them on a longer bar so that their "red" and "blue" type ends are joined. Using a compass, we show the indeterminate middle zone, as shown before for a

Experiments in Magnetism

bar magnet. But, if we now separate the two parts, in the previously neutral middle zone, suddenly two new poles appear!

MAG 4 MAGNETIZATION

a) Two large nails, *not* magnetized, are suspended next to each other on strings. Now they are put near a modern, powerful alnico magnet (strong enough that a student cannot remove the yoke by pulling straight up). We allow the nail points to jump onto the magnet. Then we observe how the nail heads move apart from each other (Figure 26b). Experiment

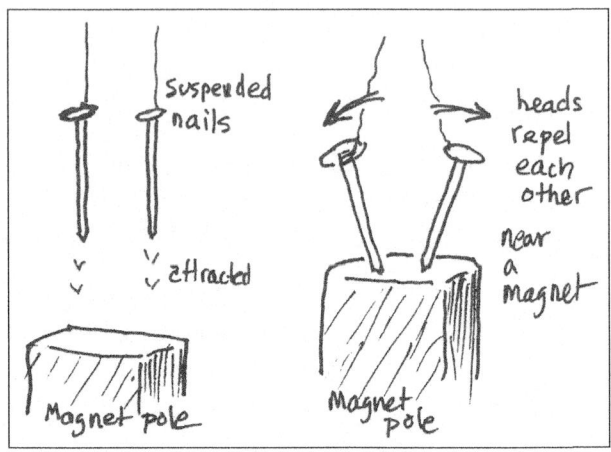

Figure 26b. Proximity magnetizing

a bit, and observe carefully, then it becomes clear. [The nails have become little magnets in the vicinity of the big magnet. Both heads are now are the same kind of pole as the big magnet pole to which the points are attached; thus the heads repel each other.]

b) CHAIN OF NAILS. Using a series of nails laid next to each other, we carefully try to pick up one end so the rest come along. When we pull off the uppermost, the rest fall off. Nearby the compass needle shows weak attraction and repulsion phenomena.

c) DEMAGNETIZING can be easily accomplished by heating or by pounding (Grade 6, Mag3). Also we can utilize the electromagnetic coil (Grade 8) powered by alternating current—since it constantly switches back and forth through the null point, it leaves a bar in its center space with no net magnetization.

MAG 5 FIELD

Cover a large, thin board with a sheet of paper and evenly sprinkle iron filings (fine, weakly magnetizable iron powder obtained from a science supply house). Bring a large, right-angle permanent horseshoe magnet underneath. Near the poles, a pattern immediately arises: Over the pole areas, the iron filings pile up forming a "fur coat." Then, positioning cards at various distances above the magnet and paper sheet, we can sprinkle each evenly using a filings "shaker" made from a small box like a salt-shaker. Thus, level by level, we build up a three-dimensional image of the field pattern, visualized using these cross-section planes. It can also be visualized when we dip a bar magnet directly into a pile of filings and let a "beard" build up on one end. However, this is not such a good idea, as the filings directly on the magnet can be removed only with some effort.

MAG 6 MAGNETIC SHIELDING

A short bar magnet (or the so-called "keeper-plate" or "yoke" from a powerful magnet) is placed on a supporting background, e.g., a book slanting toward the students. If we move a magnet beneath, then the yoke above moves also. The magnetic influence penetrates through paper, wood, glass, brass, aluminum, water, and even through our hand. But it will fail to penetrate a thick iron (steel) plate. Water's lack of shielding can be demonstrated using a shallow tray and an iron strip in a boat on a thread. The magnet beneath still moves it.

MAG 7 CONCEALED MAGNET

On top of the lab table, place any convenient, ordinary piece of marble or wood. Underneath, secretly fasten a bar magnet with masking tape. Using a compass, we allow a student to determine where north is in the classroom. But, above this bewitched surface, the compass needle becomes "confused" when it is placed at various locations. We let the students conjecture what the situation is there (an iron bar, concealed magnets, etc). We investigate the situation by determining the compass orientation at various locations on the table surface or on a sketch of the table. Thus, we arrive at a sketch of the field of the concealed magnet, and can clearly say where its "red" (north-seeking) and its "blue" (south-seeking) ends lie. We let the students find this out on their own. Only afterward do we reveal the hidden magnet to verify the solution they have worked out independently.

Grade 7 Mechanics

OVERVIEW

We recall the short sentence from Rudolf Steiner's curriculum lecture already mentioned in Volume I, in which he describes the mechanics chapter for this physics main lesson:

> And starting from there, you go over to the most important basic ideas of mechanics: the lever, crank, pulley, inclined plane, wheel/roller, Archimedian screw, helical thread, etc.

Of the four, perhaps five weeks (better three plus two) allotted to physics in 7th grade, one or at most two can be devoted to mechanics. Just the lever and its applications could occupy us for one whole week. Looking over the program indicated by Dr. Steiner, we immediately see that most topics can be presented only in general impressions and should not be studied through exact formulas or even computations [at least not in the presentations in the block itself]. The force-multiplying apparatus introduced here (the so-called simple machines) are presented in relation to bodily force and action. There is only one that, in its widespread everyday applications, is usable only through a knowledge of exact arithmetic: the balance—an application of the lever. Precision is essential: It must produce exact weight, especially as a gold scale! The other hand-operated machines have to do with approximate assessment of the force-increase, i.e., with the increase of ease, most importantly in the way they make work easier. Thus, we study the lever first, initially only for large-scale amplification of force, without mixing in any numerical calculations.[1]

I. THE LEVER

The LEVER is introduced along with the crowbar as an apparatus that produces a lifting effect. The word *lever* comes from the Old French *lecour*, to "heft" or "lift up" (Me1). The students experience how the loading, downward-pressing forces of the world surround us. In the forest trees, struggling upward against this heaviness, they can see how people were inspired to utilize timbers taken from it in order to lift up others. This also occurs when we use iron and steel—reduced from ore by means of charcoal (also a by-product of life)—as a load-bearing bar or girder. Soon the students will discover that the lever's effectiveness ("mechanical advantage") has to do with the length of the arm. The bar should be placed in such a way that the human arm receives a kind of extension. This is verified in the experiment with a long beam (Me2). The force of weights on equally long arms acts equally—as we know from the see-saw.

This is not the case, however, when one weight sits on "a longer lever." This can be formulated in words: The longer the lever-arm, the stronger the effect of the weight on it. Conversely, the shorter, the weaker the effect. Can we increase the force this way, indefinitely? This would certainly require very rigid lever-arms. But, what generally happens inside the lever? How does it produce the amplification of force? Does it all remain under the same stress? (Me3). The persistence of a central thickening shows that the rod is not under stress at the ends, where the load is applied, but the greatest stress exists at the fulcrum point. A precise investigation

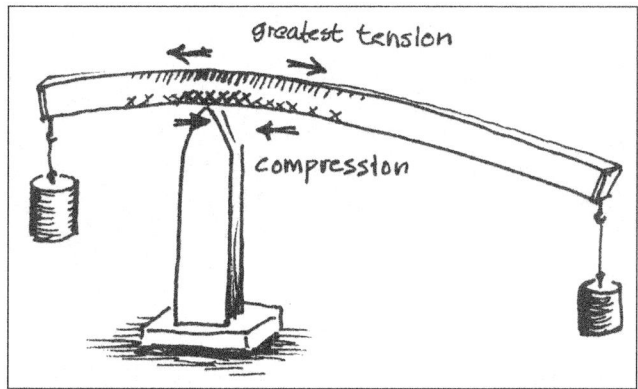

Figure 27. Forces on a 2-arm lever under load

shows that in bowing, the lever-arm undergoes extension on top and shortening underneath. Above, it experiences tension, and below compression. These forces increase toward the fulcrum and decrease toward the ends.

If, through the force of weight placed on the right end, I can hold up an equal weight on the left, then without having to alter the weight on the right, I can also hold up a much greater weight on the left if it is nearer the fulcrum. But, by altering the weight on the left (small further out, larger nearer in), I have not changed the stress inside the lever-arm near the fulcrum since the unchanged right-hand weight continues to hold the balance. This great stress within a lever-beam near the fulcrum is present as long as the lever is in action and is the same whether the weights are larger ones closer in or only smaller ones further out. The localized stress in the lever-arm corresponds to the total weight placed upon it. The load-force that increases stepwise as weights are placed closer in from the ends of the lever-arm toward the fulcrum is expressed as an inner increase of stress along the spine/backbone of the lever.

Mechanics

A feeling-participation in such inner forces is not speculative ornamentation nor should they be described only by a mathematically clear lever-law; rather it marks the actual pedagogical goal of this topic. The learning of formulas and calculation with numerical values and, above all, the study of pure motion (phoronomy) are abstractions of the lower senses (mass, number, position); they educate an intellect separated from experience. In contrast, a feeling-participation in the inner stress of the actual lever leads to a concrete mode of study that includes the actual experiences of all the senses.

The student asks himself, "How would I feel inside if I were the beam onto which these external forces are applied?" With our will experiencing the forces in this way, our senses for movement, balance and inner-life are awakened to perception and are connected to the higher senses since we also orient ourselves to the bending (sight) or the creaking or cracking (hearing). And, if the beam is now loaded to bursting, this arouses our feeling-participation even more. This participation in feeling can be characterized after the experiment and finally worked through with concepts (e.g., the localized stresses in the wood). The whole person is thereby brought into play; the illumination goes stepwise from the willing, through feeling, to the discernment of thinking. In this second seven-year period of the youngsters' lives, it is just such a participation through feeling-experiences, then differentiated through the details of the observations, which is most deeply educative.

Therefore, it would be misplaced to present sensational experiments with mighty, rending forces, just to abstract them to externals, i.e., phoronomical formulas (of pure movement) such as the lever-law or mere calculation of numerical equivalents. This really would amount to pulling a thin layer of intellectualism over a chaotic will; it is amusement and abstraction in place of a penetration suited to feeling. "If you don't feel it, you won't pursue it."[2] If a real feeling-connection with each actual place along the lever-arm is experienced and considered, then we can confidently pluck out our theme from the pedagogy. Certainly, the teacher will eventually inform them about the numerical equivalents of the lever-law and describe examples. But, in general, we will first present a number of experiments and then—after having established the experiential connection more securely—exercise a little intellectual mastery to unite the awakening forces of intellect with the exertions of the will.

With this, we move on to the BALANCE, where the inner experience of the lever-arms is equal and depends only on the weights placed on them at the ends (Me4). Then, we could perhaps present the difference between the single-arm and double-arm balance. (In a previous discussion, we referred mostly to the double-arm balance). From these considerations, all sorts of applications of the lever can be found. For example, with the street fruit-vendors, e.g., in Italy, one often runs into the old Roman double-pan balance (two-armed scale). Balances of this type are still in use in the form of the hunter's scale or the farmer's animal scale. These can easily accommodate the weight of a student and are quite useful for demonstrations (Me5).

What our experiment on the thickness of a lever-arm teaches us is visible in an artistic form in antique balances, also in the drawing *Melancholia* by Albrecht Dürer. Another form is the bottle opener, which is a single-armed lever [also described as a "Class II" lever = the lift is located on the same arm as the load, though with a longer arm]. The nutcracker and the crowbar are the same type of lever. Pliers are like a double-arm double lever: two levers working in opposition [a "Class I" lever = load and lift are separated by fulcrum]. The spade is also a double-arm lever, as it is commonly used. The long paper shears provide a translation (increase) of force from the hands—deep in the jaws at the start of the stroke—for cutting thick materials, while at the end of the stroke—at the tips—they produce a reverse translation giving more rapid action (slow, smaller finger motions produce greater, faster motion); so, the paper can be cut quickly before it folds out of the way of the cutting edges. A fishing pole is a single-arm lever [a "Class III" lever = the lift is closer to the fulcrum than the load]. Its usefulness is in its long reach rather than its mechanical advantage or lifting capacity.

With the single-arm lever, we lift in the direction of the desired movement of the load; with the double-arm lever, the reverse (press down, when we want to raise something).[3]

II. WHEEL AND AXLE

With the lever, we are dealing with forces applied to an almost totally mobile bar. Next we go into machines connecting such forces to a bar that is firmly mounted and can only rotate (AXLE AND WHEEL). In the Tibetan culture, everyday application of the wheel and axle was avoided until about two decades ago. While the advantage of a rolling wheel, e.g., on an ox-cart, over a simple skid is easy to comprehend, it is not so easy to grasp the forces acting from the axle out into the wheel. What takes place within the axle, which somehow carries the force to the wheel? As a simplification, we initially imagine an axle with only one transverse arm on each end:

Mechanics

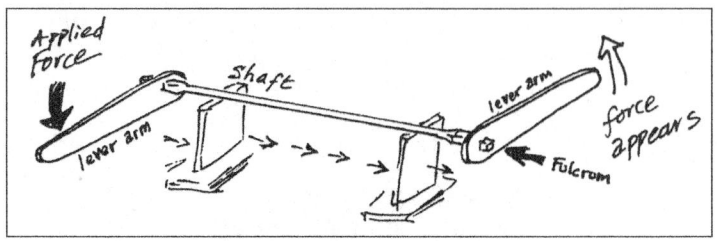

Figure 28. Axle with arms

The shaft (axle) of our axial-lever forms a separated fulcrum for two half-lever arms. The rotary force that acts inwardly on such a shaft and deforms it, leads usable force (applied at one lever-arm) from one end to the other. In contrast to the lever, the axle is under stress (torsion) along its entire length, but not equally over its cross-section. On its outer surface, in the mantle-layer, it is mostly under tangential or rotational stress (torsion). Toward the center, these forces decrease. Therefore, in order to conserve material, one can use a tube without losing the stress-carrying capacity. Thus, the drive shafts of vehicles, designed for the force they must transmit, are always a hollow cylinder.

For our next experiment (Me6), we use our special wooden axle with arms (cranks) of various lengths attached. On the one end, the stress vanishes through the square arbor into the axle and reappears on the other side (Me6b). We see that the most stress exists in the periphery of the spindle, not in its core (at least the torsion is visible by the indicator stripe). If we live into the inner conditions of the axle in feelings, then we can get a sense for the inner twisting—in contrast to the bending of the lever. With this experience of torsion, we realize the most important pedagogical point of the mechanics of the axle. Next we move on to a consideration of the wheel & axle and finally a numerical treatment of it (Me7).

Imagine a well. A chain with a hundredweight bucket (100 lb) leads upward and winds around a round beam. The beam is rotated by a wooden wheel, over which runs a cable. A fragile, old woman draws on the end of the cable. How does she raise the bucket? Apparently the situation is this: The larger the wheel, the smaller the weight or force necessary on her end. In the wheel we again recognize the axial-lever arrangement; the spoke of the wheel (radius) is the lever arm. At equilibrium (when motionless), we can imagine the cable as glued to the wheel at the point of contact, or else attached to a hook there. From the hook to the center of the axle is a lever-arm (the radius). While the wheel turns, this point shifts along its circumference; however the radial lever arm remains the same length. And, if we imagine the area the cable contacts along the perimeter of the wheel as dividing the force—as the cable is clamped by friction along a long contact surface—the force is spread out over a larger area. But, most important, it is always connected along a lever arm (radius) of the same length. All these partial levers (radii) taken together produce the lever-action seen above.

III. ABOUT FIXED PULLEYS AND BLOCK & TACKLE

A FIXED PULLEY can be used like a balance. If one knows the weight of a student, we can see how it is balanced from either side of a fixed pulley by an equal mass hung on the other side (Me8). This pulley is often used in construction sites and in lofts. It permits a load below to be raised with a downward pull and also allows the bricklayer or worker to keep his position up above. It doesn't reduce the force needed, it only reverses the direction of pull.

The MOVABLE PULLEY, in contrast, already produces a reduction in the force required (mechanical advantage) because the worker has to lift only half the load (Figure 29). As the pulley can rotate on an axle, the pull on cable segments A and B is equalized. Cable segment A lifts the pulley along with the load, in cooperation with segment B. If we substitute a person for the hook, then that person would have to lift just as much weight as with a single pulley.

Because the force in segments A and B both carry the weight, each segment thus carries only half as great a force as the downward force at the hook. This arrangement, considering the segment from the worker to the movable pulley, produces a halving of the load (and a doubling of the work or motion). As we already saw with the lever, this increase in advantage comes at the cost of motion: The worker has to be only half as strong as with no

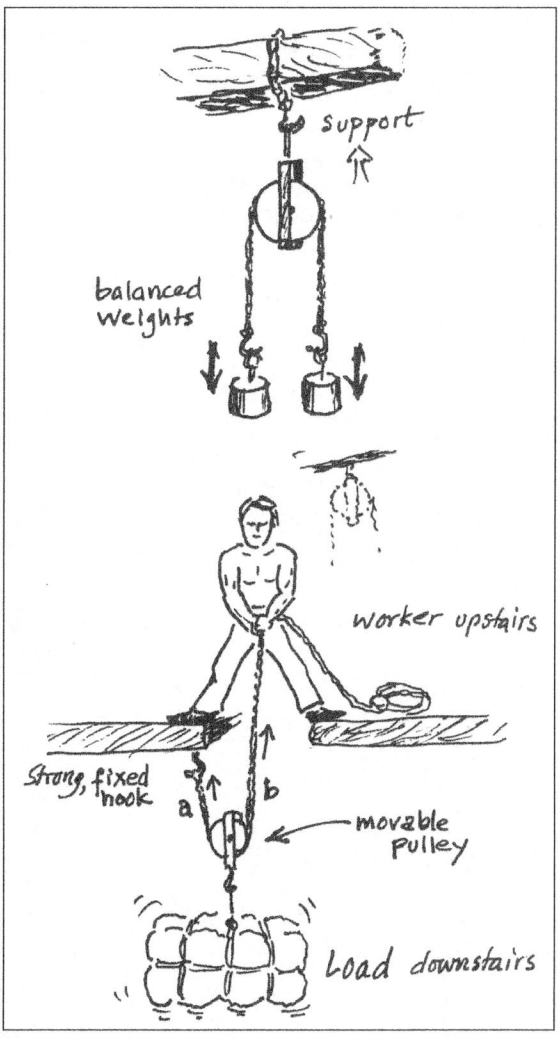

Figure 29. Movable pulley: reduction of required lift force

pulley, but he has to pull twice as far. He must lift twice as much cable than if he raised the load directly with a cable alone. Apparently, he has to also pull up the rope on the hook side, in place of a second person.

Now, along with the advantage produced by having a lower (movable) pulley, we can run the lift cable over a second pulley (fixed) located up where the worker stands, e.g., a fixed pulley fastened to his ceiling. Now, the man can walk to the side and very easily pull downward or diagonally (Me9).

The BLOCK & TACKLE is simply a series of equal number of fixed and

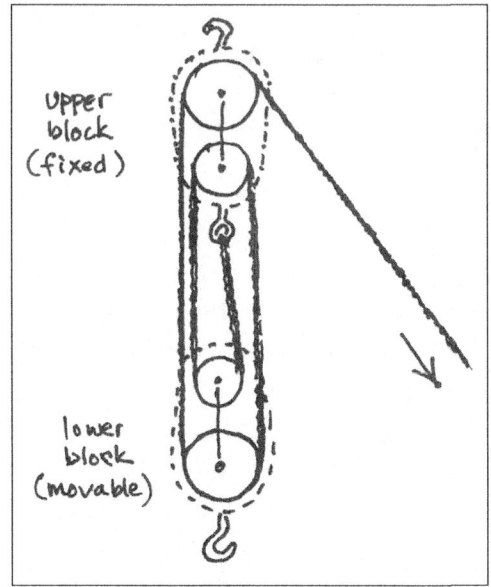

Figure 30a. Two-pulley

movable pulleys, each mounted in their own ash-wood retaining blocks, termed a "pulley-blocks" by sailors. The upper block (Figure 30b) contains the fixed pulleys, the lower one contains the movable ones. Between upper and lower blocks we count four cable segments. They are connected to each other and to the lifting end of the cable only via the movable pulleys, and they each exert the same pull.

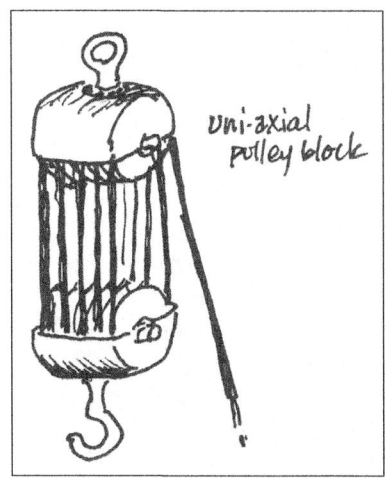

Figure 30b. Block & tackle

The pull, which the worker exerts on the end of the cable, acts fourfold on the lower block; it can carry a load four times larger than a single cable. Naturally, in raising the load, the worker has to shorten all four cable segments and has to pull up a fourfold length of cable through the upper block (Me10). In construction cranes, which, e.g., lift prefabricated building pieces into place, at the lift hook block, we can usually find a block with many more cables since these are not twisted as easily as a single one. Nevertheless, unlike the antique block and tackle still used today for raising logs, in the newer pulleys, the pulley wheels all sit on a common shaft next to each other and not one below the other (Me4b). But, the older arrangement (Me4a) is better for classroom demonstration, since the cable motion is more visible.

With two pulleys, we form a machine with increases the force twofold. But with four pulleys, fourfold. How many pulleys are needed to increase it threefold? (See Figure 31).

IV. INCLINED PLANE, SCREW & WEDGE

While the lever, wheel and pulley lead into numerical relationships, to show such numerical relationships with the INCLINED PLANE would lead us too far afield; so, this is deferred to the treatment of force-vector parallelograms in grade 10. Instead of these considerations, it may suffice to set up an inclined plane, and pull up a large weight along an inclined plane, using a small wheeled cart, with a pulley at the top to reverse the lift force; you can do this with different inclinations.

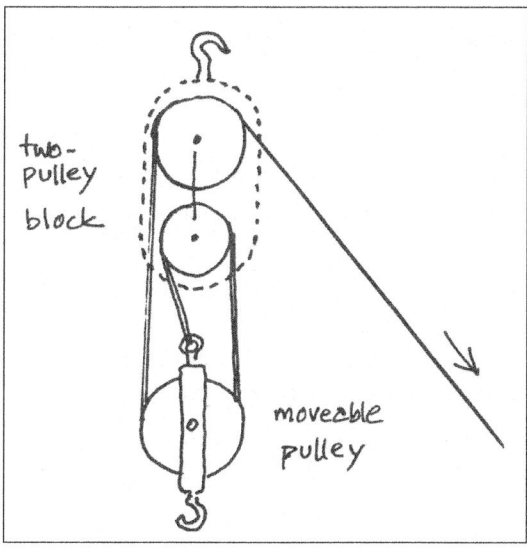

Figure 31. Three-fold force increase

Result: As the travel surface gets steeper and finally vertical, we can even remove it—the load and the force required have become equal and the surface no longer carries any of the load. In contrast, the more horizontal the travel surface, the smaller the pull needed (almost nothing with a frictionless cart on a horizontal surface). But the load gains much less in height at the same time.

Also, the more horizontal the inclined plane, the smaller the forward force required to overcome a large downward force (the weight of a massive load) and even raise it—if only a small distance. An example would be a massive capstone for a burial vault, a massive load which can be moved into place only by using an inclined plane. Often, the forward motion is made easier by using rollers in addition to the plane.

The force advantage of the inclined plane is also put to use in the SCREW. We wind the inclined plane around a cylinder, forming a winding spiral path. A large force driving the screw-point forward is thus produced from a smaller force rotating the screw. The so-called Archimedian water screw can in this way lift a much greater weight of water using only a small force to turn the apparatus.

A much simpler application of the inclined plane is seen in the axe head or in a log splitter; they are examples of the WEDGE which also makes use of the mechanical advantage produced by the diagonal force of the inclined plane.

Experiments in Mechanics

ME 1 CROWBAR (PRY-BAR)

A 50 kilo (110 lb) stone block is set up with one end on a thin wooden shim so that the sharp end of a crowbar (so-called pry bar or crate-opener) can be inserted underneath. What the strongest student can hardly do with his fingers, a small girl can easily do with the crowbar—even if many students sit on the block. If the room does not have a hard floor under the block, place a steel plate underneath the supporting point of the crowbar (to prevent damaging a wooden floor).

ME 2 LEVER

Have two students sit on a small bench. Set a beam (4" x 4") on top of a sufficiently large wood block/stump so that one end is positioned just under the bench. We investigate the lever lengths needed for a third student—using his weight—to lift the first two along with their bench. Provided all three weigh about the same, the length ratio (bench-fulcrum):(fulcrum-lifter) should come out 1:2. Using different lever arm lengths, the "lifter" can only lift one student; now the ratio of lengths should be about 1:1.

ME 3 CANTILEVER BRIDGE

We glue laminate boards into assemblies as below (Figure 32), which all have the same overall length but different arrangements of laminated pieces.

A half-round piece of wood is nailed to support block as a fulcrum. Assembly b will withstand a load on both ends which will cause assembly c to snap, while assembly a makes ominous splitting sounds but only cracks slowly. (The greatest stress is located in the center, near the fulcrum, NOT at the ends where the loads are placed.)

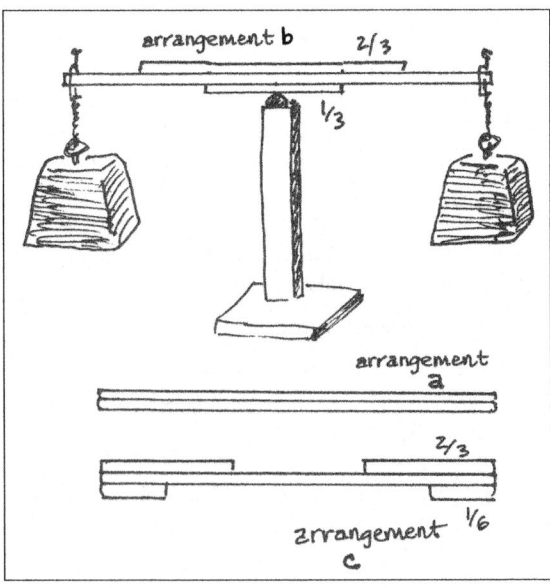

Figure 32. Laminated cantilever

ME 4 BALANCE

Set up a balance beam about 122 cm (48") exactly on its center point—i.e., in equilibrium—on top of a fixed support beam; the balance beam has about 12 hooks placed every 5 cm (2"). The various balanced arrangements can be demonstrated or verified with equal-sized weights hung on various hooks, e.g., 2 x 9 = 2 x 3 + 2 x 6 (2 weights at lever arm-length 9 just hold in balance 2 weights at position 3 and 2 weights at position 6.

ME 5 ANTIQUE FARMER'S BALANCE

We can actually weigh a student with this heavy-duty two-pan balance. It depends only on how reliably the support is fastened into the ceiling or rafters. The seat for the student is borrowed from the block & tackle apparatus. Use various weights in the other pan to measure the weight.

ME 6 CRANK (AXIAL LEVER)

Our wooden "axle lever" with torsion-resistant ash-wood shaft, has square ends to attach lever-arms on either end (Figure 33). Initially, on one end we put both a large and a small arm. So we have an ordinary two-armed lever with the shaft forming a fulcrum. In a contest, the student pushing on the long arm will beat the student pushing on the short arm (as we know well from Me1 and Me2).

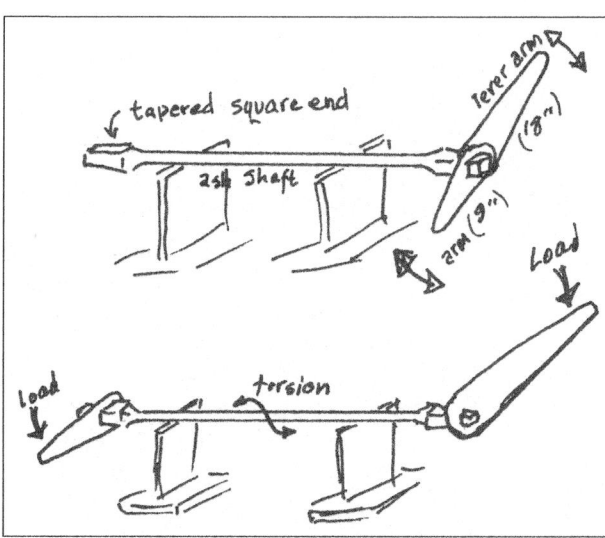

Figure 33. Axle with lever-arms

Next, we make use of both ends of the axle-shaft when we apply force to the separated levers. The ash shaft now undergoes a torsional strain, and both levers move a bit against each other. (A black stripe along the shaft shows the twisting along its length.)

Experiments in Mechanics

ME 7 WHEEL OR DRUM-WINDLASS (STEP-PULLEY)

A square-shaped metal shaft has various size circular discs mounted on the ends (a step-pulley), around which we wrap cords (in opposite directions) with weights on the ends. First we show the nearly weightless equilibrium existing between equal weights on equally large wheels. Using different-sized discs, we observe the different falling speeds of the weights. Then get them in equilibrium by adding weights to one cord (the one attached to the smaller wheel) so that the relationship of the weights-to-wheel-size is clear. [Note: The tiniest deviation from perfect roundness of the wheels, perhaps due to incomplete sanding of wooden wheels as much as from variations in the weight of the twine, will cause a small discrepancy in weight to show up at equilibrium, a false equilibrium weight. Since we want to find the ideal weight, and avoid such errors, we use a metal shaft, perhaps mounted in oiled bearings to eliminate friction of rotation as much as possible, and use very carefully machined winding drums or wheels.]

ME 8 FIXED PULLEY

Our forced pulley apparatus, capable of handling 200 kg (440 lbs) is hung from an equally strong ceiling hook (perhaps in the gymnasium). A cable (hemp, at least 8 mm thick) is threaded over it. Two students hang on each end of the cable and hover approximately in equilibrium; it is easy to tell which is slightly heavier. Weights hung on one side will measure the student's weight as with the balance above.

ME 9 MOVABLE PULLEY

We need at least two ceiling hooks of 100 kg (220 lbs) capacity and positioned about 1.5 times the pulley diameter apart (See Figure 34).

One student can hang on the movable pulley. Assess the force on the cable while he dangles there (it must be good for half his weight). Then lift him up a bit and observe or measure with a meter stick how much higher he is and also how

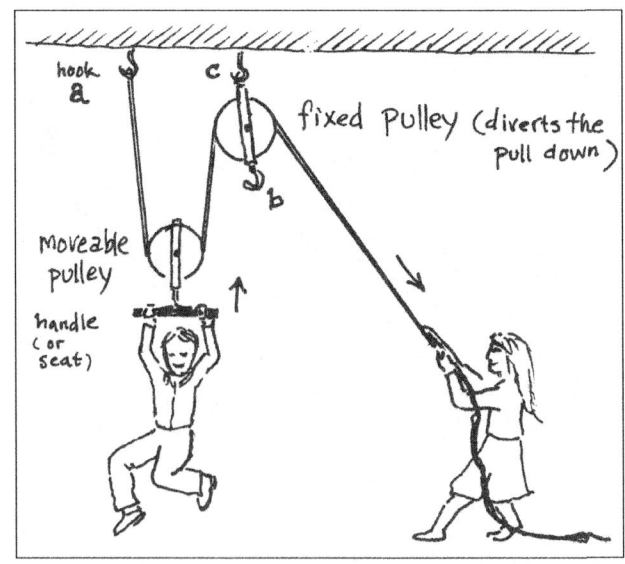

Figure 34. Movable pulley

much cable we have drawn down (twice as much). Now shift the cable attachment from the ceiling hook a to the hook b on the enclosing guard of the diverting pulley (fixed to the ceiling); nothing is changed, but we have the beginnings of a block and tackle. Then, we need only one ceiling hook, namely c.

ME 10 BLOCK & TACKLE

A four- or sixfold block & tackle is used to raise a student sitting on a board or seat attached to the lower block. Even the weakest student can pull him up "with one finger" but will have to move six times the length of cable. The rider can even pull himself up with ease. (In this case—with six rollers—he only has to exert a force a bit less than 1/6 his weight, due to his being made lighter by his own upward pull on the loose end of the cable. One-seventh his weight is now supported on the cable, so 6/7 rests on the seat. This is mentioned mainly for the teacher's interest.) It is interesting to observe how fast each segment of cable runs over its pulley roller: The outer ones—nearer the loose end—run faster than the inner segments.

ME 11 SCREW & WEDGE

If there is time, we can easily demonstrate the inclined plane using a heavy cinder block on a cart with roller-skate wheels. Pull the loaded cart up a runway made from a wooden board (9" by 4'), positioned at various angles to the horizontal and with a fixed pulley at the top. Thread a cord from the cart over the pulley so we can assess the force needed to move the cart at various inclinations.

A visual demonstration of transforming the inclined plane into a screw is easily made using a triangle of cardboard (the hypotenuse represents the inclined plane) which we wind around a cylinder and fix in place with tacks. Now the diagonal edge shows up as a winding spiral up the cylinder. Rotation of the cylinder is now equivalent to the horizontal motion produced by the cord on the cart in the above set-up. Discuss various applications of the screw: tightening thread in a c-clamp, worm gear, hand auger (drill brace) from woodwork, etc.

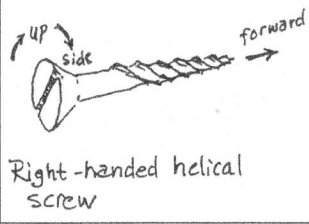

Endnotes

OVERVIEW

1. See also the various writings of Michael Polanyi on the theory of science.

2. M. Wagenschein, "Rettet die Phänomene" ["Saving the Phenomena"] in *Der mathematische und naturwissen-schaftliche Unterricht* [*Math & Science Teaching*], Vol. 30, 1977, pp. 29–137.

ACOUSTICS

1. In the SI metric system; previously it was called cycles per second, or c.p.s.

2. This unchanging, sound -while it lasts-sounds very "electronic" or "high-tech," which gives us a clue to what is lacking from such 'artificial' sounds: their complex variation (the timbre) and a 'colorful' spectrum of overtones.

OPTICS

1. "On the Value of a Thing," the mirror image in water, A. Solzhenitsyn, 1970.

2. See the mineralogy essay by F. Benesch on the significance of precious gems and the meaning of *precious* in connection with being ideal, perfect.

3. Note that its glazed surface is produced at high heat in a kiln, which brings the glaze into a semi-fluid state. The partial mirroring is not, then, surprising.

4. Faust, Witch's Kitchen scene, Ln 2429.

5. See discussion of shadows under Grade 6, Section III.3

6. In grade 7, avoid a skeptical view of the world so the development of thinking judgment of the children will not be overloaded. This poem is used here only to express the *character* of a mirror-view.

7. The labels, i.e., we now have to turn around to see the same configuration of labels as before on the coordinate cross.

8. Thanks to Dr. Georg Maier, Dornach, for the suggestion of conceiving the mirror-law starting from the yonder-twin.

9. For this clarification, I am thankful to Hermann Muller, Wanne-Eichel (Hibernia School).

10. See, "Isfahan, Mirror of Paradise" by Henri Stierlen, Atlantis Pub. Zurich, 1976.

11. Faculty Conferences, 29 April, 1924.

12. Angle of *incidence* is the angle the turtle's line-of-sight makes to the mirror; angle of *reflection* is the angle the observer's line-of-sight makes to the mirror.

13. For further discussion of these ideas for parents, see articles by Dr. Mackensen in *Erziehungskunst* [*Education as an Art*], 5/78 & 7/78.

HEAT

1   The accurate value is, of course, 98.6°F.

2   Note that the divisions of the Fahrenheit scale are finer (5/9) than the Celsius scale used in the laboratory. It can thus express a change in warmth which is just perceptible (1°F) and has a human-scale character. The Celsius scale is deemed useful since it divides two physio-chemical phenomena (freezing & boiling) into even decimal parts (0° and 100° respectively).

3   Use tubing, preferably of Pyrex, to avoid thermal shattering.

ELECTRICITY

1   Goethe, *History of My Botanic Studies*.

2   Actually, technicians generally imagine current as electrons moving at the speed of light in the opposite direction. However, as discussed earlier (Grade 6, Light, II. "Seeing…"), this "velocity" is not a primary phenomenon but a theoretically derived one, used to explain other, observed phenomena. Thus, at least to our ordinary perceptions (phenomenologically), electricity acts virtually instantaneously. These puzzles are taken up in further detail in grade 11.

3   This can be established as follows (at least in thought): Glass rubbed with silk can be shown to be positive with reference to the rubbing cloth (and also therefore, in relation to our body and to the earth, since we are standing on the floor). The charge is positive, i.e., the opposite type from that on amber or ebonite rubbed with wool; when suspended on a thread near an ebonite rod rubbed with felt, the rods are attracted to each other. If we make a battery of 1000 cells, connected as in Figure 19 (top), ground the lowest zinc plate to the earth and connect the uppermost copper plate to a metal sphere. The sphere will repel our glass rod; therefore, both are positive.

MAGNETISM

1   Goethe: *Poetry and Truth*, "Book Four"

2   To avoid this confusion, we term the north-seeking pole as having "red" magnetism and the south-seeking pole as "blue."

MECHANICS

1   A rich description of the mechanics in Grade 7 is provided by Walter Kraut in *Erziehungskunst [Education as an Art]*, 11 p. 424 (1972). He presents a somewhat different, more phoronomical goal (calculations without units). Moreover, he takes up additional topics, e.g., translation, which will be consciously omitted here.

2   Goethe, *Faust*

3   There are numerous levers at work in the body. But, since the metabolic-limb system of children is only just beginning to really physically penetrate their sinews at this age, we defer a study of the levers in the architecture of bone and musculature until the anatomy block in grade 8.

Selected Bibliography

Duit, Reinders, et al. "Everyday Concepts and Natural-Science Teaching," 1981.

Feyerabend, Paul. *Knowledge for Free Human Beings*. Suhrkamp Publ., 1981, p. 18.

Fischer-Wasels, Horst. "Non vitae, sed scholae discimus?" in: *Die Hohre Schule [The Upper School]*, 9/78, p. 339.

Fromm, Erich. *To Have or To Be*. Basic Books, 1976.

Heisenberg, Werner. "Physics and Philosophy," 1959.

Jung, Walter. "Phenomena, Concepts, Theories: 3 Theses of Scientific Theory and Didactics in Physics," in *Physics Teaching*, Vol. 16, May 1982.

Jung, Walter, see Duit, et al., 1981.

Kinzel, Hebnut. "Scientific Knowledge and Human Experience," in *Biologie in Unserer Zeit [Modern Biology]*, Vol. 9 No. 4, pg. 112 (1979).

Schumacher, E.F. "A Guide for the Perplexed," Harper and Row, 1977, p. 166.

Silkenbeumer, Rainer. "Model Schools – School Models," Hannover, 1981.

Steiner, Rudolf. *Philosophy of Freedom*. 1918, 1984.

Weizsacker, Carl F. *The Unity of Nature*. Hanser Publ., 1971, p. 244.

Wieland, Wolfgang. "Possibilities and Boundaries in Scientific Theory" in *Angewandte Chemie [Unified Chemistry]*, Vol. 93, p. 633, 1981.

Printed in Dunstable, United Kingdom